The Carlsberg Laboratory
1876|1976

J. C. Jacobsen (1811–1887) the founder of the Carlsberg Laboratory.
Painted 1886 in the winter garden of his villa in the grounds of the brewery.

The Carlsberg Laboratory
1876|1976

PUBLISHED BY THE CARLSBERG FOUNDATION

Edited by H. Holter and K. Max Møller

RHODOS INTERNATIONAL SCIENCE AND ART PUBLISHERS, 1976

THE CARLSBERG LABORATORY

LAY OUT AND COVER: MICHAEL MALLING

ISBN 87-749-6539-5

PRODUCTION: RHODOS PUBLISHING HOUSE, COPENHAGEN
PRINT: BORDING GRAFIC A/S

Contents

Editor's preface.

When I was given the task of compiling the material for a history of the Carlsberg Laboratory, it was pointed out to me that the book was intended to be read not only by scientists but also by the »general public« – whatever that may mean.

This posed the problem of the use of technical language. The Carlsberg Laboratory is a scientific institution and its history is closely linked with the development of certain branches of natural science. Therefore, if the history is to be presented at all adequately, it has to be presented by scientists. In most chapters an effort has been made to avoid an excess of technical jargon, but a certain amount of specific terminology has been unavoidable, if the book was to be more than a superficial array of facts and anecdotes.

On the other hand the Carlsberg Laboratory has in many ways been a remarkable institution right from its start, and it has harboured some remarkable people. It is therefore hoped that the »general reader« will be willing to put up with passages of technical information for the sake of the institution's historical and human interest.

This book consists of contributions from many people, some of them now dead. Notes about the authors of the various chapters and about others who have contributed in different ways will be found in chapter 13. To all who have so readily consented to join the team, I am extremely grateful. But I should like to express my special gratitude to my co-editor, Knud Max Møller who has not only compiled, edited and annotated (chapter 14) the illustrations but has also with great effort tried to ensure that the data given are historically correct.

Special thanks are also due to Mrs. Bente Gad and Mrs. Lilli Nielsen, both from the office of the Carlsberg Foundation, and Mrs. Lillian Olsen and Mrs. Eva Mahler from the Biological Institute of the Carlsberg Foundation, who prepared the index.

Finally, some points of terminology should be mentioned. According to present usage the two departments of the Carlsberg Laboratory ought to be named Department of Chemistry and Department of Physiology. However, since for close on hundred years they have been called Chemical and Physiological Department by the people who worked there, the old names have been used in all the historical chapters. Not until chapter 12, which deals with the Laboratory of to-day, have the present designations been adopted.

Besides this, various other inconsistencies will be found throughout the text. Some of these have undoubtedly simply escaped the editor's notice but others were retained on purpose in order to preserve the intentions of the various authors. It was felt that in a book of this kind complete homogeneity of style was neither attainable nor desirable.

In their choice of illustrations, the editors have been guided by considerations of historical and human interest rather than care for photographic perfection.

H. HOLTER

I

Introduction

by *H. Holter*

When it was decided by the Carlsberg Foundation that one of the publications marking the Foundation's centenary should deal with the history of the Carlsberg Laboratory, a good deal of thought was given to the question of the scope of such a book. The truly relevant facts about the history of a scientific laboratory are of course laid down in the scientific papers that have been issued from it. However, it was felt that a story based on these facts alone would scarcely make a suitable centenary publication. In order to do justice to the festive occasion we thought it desirable to produce a book describing what actually happened in the laboratory during that century, to include the human aspects of its period of growth and achievement, and the interplay between the laboratory and the world outside.

It was therefore decided to make the scientific accomplishment of the laboratory only one of the aspects to be treated. In addition the personalities of the men involved should become discernible as well as the background of their work: the way the laboratory came into being, the organization and scope given to it by its founder, the brewer J. C. Jacobsen, the way in which this frame was filled by the people who worked there, and the way in which their achievements were received.

The Carlsberg Laboratory has always been a rather small institution, consisting of only two departments, the Chemical and the Physiological, and a small staff. In comparison to its size, however, its role in science has been quite remarkable.

This seems to be due mainly to two sets of circumstances. The prime component, and indeed the one indispensable condition for the flourishing of any scientific institution, is the quality of the people who direct the work. In this respect the laboratory was very lucky: several of the directors turned out to be truly great scientists.

The other component was more specific for the Carlsberg Laboratory. It consisted in its privileged status as a pure research institute at a time where such institutes were rare indeed. The laboratory was backed by a Foundation which was unique in its organization, and which was amply endowed by the money earned in the brewery. Thanks to these assets, the Foundation was able to practise a remarkably liberal interpretation of the scope of the laboratory's work.

These were indeed fortunate conditions, and they resulted not only in good research but also, during long periods, in a rather special atmosphere of tolerance and security, an atmosphere which was felt by all who came to work in the laboratory and which contributed much to the laboratory's popularity and to its achievements.

There is no obvious recipe to follow in compiling the story of an institution of this type. So, in consideration of the material available, it was decided to let the existing biographies of the heads of the laboratory's two departments form the backbone of the book. This was done in consideration of the fact that the biographers personally knew the men they were writing about and had done their very best. It was scarcely to be expected that a better result would be obtained by duplicating their efforts many years later.

The question is, however: Is it permissible to assume that the story of the men directing the work is in fact also the story of the laboratory? This assumption would certainly not be justified in the case of many other institutions, especially more modern ones, but the Carlsberg Laboratory is, we believe, a special case. This is due in part to the force of the personalities involved but also – and to a considerable degree – to the patriarchal form of organisation chosen by the laboratory's founder. According to the directions laid down by the founder, the head of the department certainly was the center of the department's activities, assisted by an amazingly small staff which originally consisted of only one assistant and one janitor. Although the size of the staff gradually increased during later years, the total number of co-workers permanently associated with the laboratory remained low compared with other laboratories.

This small nucleus of permanent personnel was, however, considerably increased during later years by a flock of visitors, young scientists from Denmark and overseas, who usually joined the working program of the department and who by their enthusiastic efforts contributed very substantially to the scientific output of the laboratory. It is undoubtedly to a large degree due to them that the danger of stagnation, so often a real threat for a small and close-knit team around one older man, was happily avoided. The contributions of both kinds of collaborators, staff and visitors, to the scientific achievements of the laboratory will be covered in the chapters about the people with whom they collaborated.

Although the biographical matter thus forms the skeleton of the book, no adequate picture of the whole history of the laboratory could be expected to emerge from a succession of biographies alone. Therefore a number of chapters have been added which deal with the laboratory as a whole.

The first of these chapters (chapter 2) will describe the origins of the laboratory, its background in the Carlsberg Brewery and in the personality of its founder. J. C. Jacobsen lived in L. Pasteur's and H. C. Ørsted's time and was deeply impressed by the possibilities of the impact of science upon human civilization. He was a true idealist but, luckily, he was also a man of action who manifested his idealism in creating the laboratory and the Carlsberg Foundation.

The subsequent history of the laboratory will, as already mentioned, be depicted in a series of biographical evaluations of the heads of its departments (chapters 3–4). From these, it is hoped, will emerge a picture of the main periods in the life of the institution: The period of ascent under J. Kjeldahl, Emil Chr. Hansen and the first years of S. P. L. Sørensen; then the period of fulfillment under Sørensen, Johs. Schmidt, K. Linderstrøm-Lang and Ø. Winge. This was the period of great achievements and international fame. Finally, the period of the recent years during which, owing to the explosive world-wide expansion of research, the Carlsberg Laboratory met with more competition from similar institutions than it had known in its earlier days. Moreover, this period brought the beginning of anxiety about the mounting demand for costly equipment and increased staff. The resulting financial stress led to the most recent development in the history of the laboratory and the solution of its financial problems by its direct association with the Carlsberg Breweries. This last period will be dealt with in a special chapter (11).

However, it must be realized that the present close association with the brewery is only an elaboration of bonds which have been a special feature of the laboratory's life all through its history. The links between brewery and laboratory were in fact such an important factor in the history of the laboratory, right from the beginning, that they also will be treated in a special chapter (10).

Another important factor for the special atmosphere of the laboratory has been the role played by the Board of trustees, a small committee appointed by the Danish Academy of Sciences and Letters. The relations between the trustees and the departments of the laboratory were not without their special problems, some of them rather illustrative of life at the laboratory through the various periods mentioned. This aspect of the laboratory's history will also be treated in its own chapter (6).

With the chapters mentioned so far, the main features of the story of the Carlsberg Laboratory should be covered from the historian's point of view. In order to give a full picture of life at the laboratory, we felt that the role of the visitors from abroad, already briefly mentioned, ought to be dealt with more specifically. However, their number is so large that it was out of question to attempt anything approaching an exhaustive chronicle of their activities. Instead, we have asked some of the old visitors and friends to give us some glimpses of life at the laboratory as seen through their eyes, and also some reflections upon the threads which, thanks to their own activities and experiences, were woven between the laboratory and the international scientific community (chapter 8).

Another feature of the laboratory's international relations is the role that the laboratory has played in the exploration of the seas during the days of Johs. Schmidt. Although most of these activities went on outside the laboratory, on expeditions financed by the Carlsberg Foundation, the results seem to justify a chapter of their own (chapter 9).

The work will conclude with an evaluation of the present: the Carlsberg Laboratory 1976 (chapter 12).

II

The origins of the laboratory and its background in the Carlsberg Brewery

by *Birger Trolle*

> …, »*dans les champs de l'observation le hasard ne favorise que les esprits préparés*«…
>
> Pasteur, á Lille le 7 décembre 1854 dans le discours d'inauguration de la Faculté nouvelle, à l'adresse d'Ørsted, physicien danois.

The Chemical and Physiological Laboratory at the Carlsberg Brewery was established with this aim: »By independent investigation to test the doctrines already furnished by science and by continued studies to develop them into as fully scientific a basis as possible for the operations of malting, brewing and fermentation«. It was founded in 1875 by the founder of the brewery, Brewer J. C. Jacobsen, and was administered directly by him until the establishment of the Carlsberg Foundation on September 25, 1876, when it was placed under a board of trustees elected by the Royal Danish Academy of Sciences and Letters according to the statutes of the Foundation.

J. C. Jacobsen's initiative was quite unique for his time. In order to understand its background one must turn to his own letter »The Carlsberg Foundation, its establishment and statutes« of September 25, 1876 to the Royal Danish Academy of Sciences and Letters.

J. C. Jacobsen writes: »In the course of time, as the efforts of scientists provide industry with more knowledge and a greater understanding of the laws of nature, there is a growing recognition of the vital necessity for such knowledge, and an increasing urge to extend it, especially in those directions which are of particular importance for the individual branches of industry.

However, since it cannot be required or expected that scientists employed in public institutions should devote themselves to any considerable extent to the numerous items of detailed information which are required by individual branches of industry, recent years have seen the establishment in several countries of special laboratories or so-called research stations, designed to provide the industry in question with more ample knowledge and a deeper insight.

These institutions have certainly been of no small service to industry, including the brewing industry, but as a rule they leave much to be desired. They almost all suffer from the disadvantage that their existence is insecure and can only be regarded as transitory, which implies frequent changes in their scientific personnel, whose preparatory studies and research are thereby interrupted before they have time to bear fruit. The aims of such institutes are often framed too narrowly, so that they do not provide opportunities for the further development, or even the preservation, of that general scientific proficiency which is essential for their management. Conversely, their terms of reference often embrace too many other duties, such as very elementary teaching (as in the so-called »Brauschulen«), which does not leave enough time or peace of mind for true scientific research, not to mention the frequent addition of purely industrial activities such as routine analyses, the issue of certificates, and similar matters.

Guided by the above considerations I have established in connection with my Carlsberg brewery a laboratory for chemical and physiological research and studies in those branches of the sciences which are particularly important for the processes of malting, brewing and fermentation, with the object not only of giving day-to-day assistance to the brewery, but also of giving the scientific staff the opportunity and resources to become qualified and active specialists in those fields which are particularly germane to the operations of brewing and the phenomena which they reveal.

I have engaged Mr. J. Kjeldahl, a polytechnic graduate, and Mr. R. Pedersen, a medical graduate, to be responsible for the chemical and physiological work in this laboratory. Further staff will be appointed later to assist the department heads, partly in the series of observations and analyses required primarily in the interests of technology, and partly in purely scientific laboratory work. These assistants will thus have the opportunity of becoming trained either as expert technicians or as scientific workers, according to their aptitude or vocation.

On the other hand, it is not my intention that the laboratory should become a teaching institution.

Since such an institute, designed for specialized studies, can only thrive if it is supported by the spirit of science and is pervaded by the light which emanates from the sciences as a whole, and since this light has for me been a source of joy and well-being, it is of vital concern to me, in order to repay some of my indebtedness, to make a contribution to the general advancement of science, especially in those directions where it appears to me that the state has not been able to afford all the necessary resources and is unlikely to do so in the future.

I allude to the following provisions: temporary grants to younger scientists whose talents and vocation make them specially suitable for subsequent appointment as university lecturers; also grants or stipends for men who are excellently equipped for scientific research or writing, and who should preferably be able to devote all their energy to these ends, without being hampered by teaching or other official duties; also travel grants to older scientists who either by repeated short visits abroad or by visits of longer duration could reap a fruitful harvest for Danish science; further, the support of scientific investigations, for example of problems whose solution is urgent at any given time, etc.

With this extended aim in view I have today established a

foundation – under the name »Carlsberg Foundation« – to which I have donated a mortgage on my property Carlsberg. However, in order that such a Foundation shall function as intended both now and in the future it requires a permanent Board of Trustees composed of men with scientific insight and ability. In this connection one necessarily thinks of the body which has always contained the most outstanding representatives of Danish science, and no doubt will continue to do so in the future; moreover, it is the only institution in this country fortunate enough to be independent of all outside scientific considerations and influences: I refer to the Royal Danish Academy of Sciences and Letters.

I therefore take the liberty of addressing myself to this esteemed Academy with the request that it will appoint a number of its members to supervise the above mentioned Foundation and to guide its activities, so as to provide for its survival now and in the future, and to ensure that it will always work for the benefit of Danish science and the honour of Denmark.«

Further, J. C. Jacobsen defines the aims of the Carlsberg Foundation:

A. To continue and extend the activities of the chemical and physiological Carlsberg Laboratory, established by me in 1875, in conformity with the aims of this institute stated below.

B. To promote the various branches of science, together with mathematics, philosophy, history and philology, in the manner stated below.

Under point A it is especially emphasized that »no result of the activities of the institute which is of theoretical or practical importance may be kept secret.«

These are J. C. Jacobsen's own words concerning the establishment of the Carlsberg Laboratory, and they speak a clear, may ask, how J. C. Jacobsen had acquired such an advanced and correct understanding of the importance of research for the development of industry. We shall try to elucidate this in the following.

J. C. Jacobsen's father was born in 1773, the only son of a farmer from Northern Jutland. His parents died in the 1790s, and the desire of their son was to go to the capital, Copenhagen. He did so in 1801 and at once obtained a job as an unskilled labourer at the »Kongens Bryghus« brewery at Langebro. He proved to be most intelligent and efficient, and he did so well that in 1817 he became a tenant brewer in the centre of the town (The Gyldenfeldtske Brewery in Knabrostræde/Brolæggerstræde). In 1826 he bought a nearby brewery of his own after he had become manager of the brewery at Langebro in 1823. He had married in 1810, and on September 2, 1811 his wife gave birth to their only child, Jacob Christian Jacobsen. At the age of sixteen, the son was employed in his father's brewery being, guided with reason and intelligence, and at his father's suggestion he began 17 years old, sometimes with his father, to attend H. C. Ørsted's popular lectures on chemistry and physics in the Society for Popularization of Physics, established in 1824.

Soon after, when the Technical University was established in Copenhagen (1829), he eagerly attended a series of lectures on natural science. He came in contact with a group of young, progressive students at the University of Copenhagen. There he attracted attention by the great store of knowledge that he had acquired, and everybody was impressed by the young »brewer chap«'s abundance of information, clear logic and many-sided interests.

When Jacobsen's father died in 1835, the son took over the brewery on behalf of his mother. At that time Copenhagen beer was dark, top fermented, so-called »White Beer« of unpleasant quality, which manifested itself at several exhibitions,

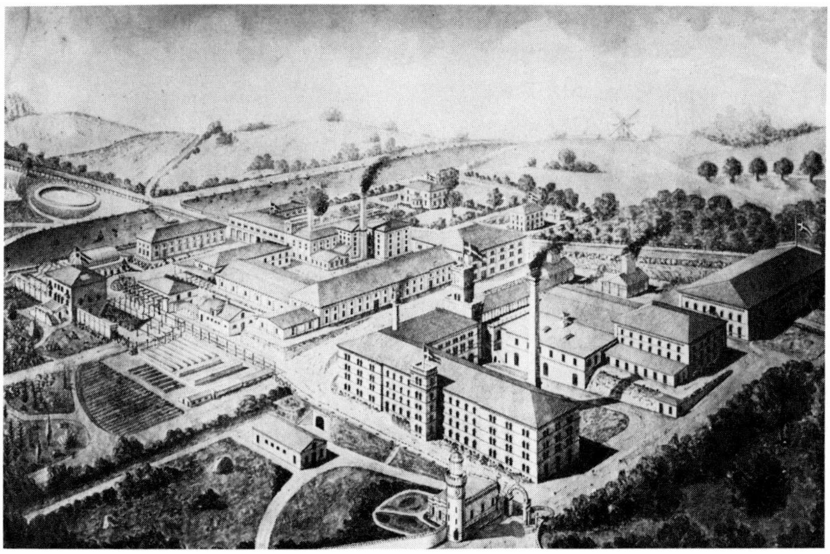

1 View of the Carlsberg Brewery, 1883.

where German bottom fermented lager beer was also presented and much praised. J. C. Jacobsen became very interested in such beer, and at an exhibition in 1844 he presented some experimentally produced lager beer which attracted much attention. At this time his mother had died, and Jacobsen inherited the brewery as well as a quite large sum of money. In order to study German beer, Jacobsen went to Hamburg and later on to Gabriel Sedlmayr's famous brewery »Zum Spaten« in Munich. With great difficulty he carried home a tin of genuine Bavarian yeast in his hat-box. He obtained the King's permission to build a cellar for cool storage in the ramparts around Copenhagen, and by 1846 Bavarian beer from »Brewer Jacobsen's cellar under the rampart« was for sale. But he wanted to go further and to build a real Bavarian lager beer brewery. In spite of warnings he went ahead with his idea. He found a suitable site in Valby (at that time a village just outside Copenhagen with good well-water), petitioned the King for permission, and in the autumn of 1847 the Carlsberg Brewery was established designed for an annual production of 4000 barrels (around 6600 hl). On November 10, the first succesful brew was started.

The standard of brewing at that time was rather low, and Jacobsen encountered many difficulties. Still he considered it a matter of honour that Carlsberg beer should be the best possible, and he spared no pains to achieve this goal. He travelled much and kept in contact with the leading brewers in Europe: Gabriel Sedlmayr in Munich, Anton Dreher in Schwechat near Vienna, Eugène Velten in Marseille, and W. Feltmann in Rotterdam. Together with these colleagues, he introduced steam into the brewery. He also introduced mashing machines and refrigerating machines utilizing C. Linde's system.

At the World Exhibition in Vienna in 1873 Jacobsen presented his bottled beer which created a sensation, and he received »The Medal of Progress« as a reward. He reported his method of heating mash with steam in the brew-house, and Carlsberg's repution as »Musteretablissement« was assured. Annual production had now passed 37.000 barrels (60.000 hl).

The natural sciences had also made great progress, and manufacturers became aware of the great importance of scientific knowledge.

The first great step was the discovery of diastase in 1833 by A. Payen and J. F. Persoz. Next came C. Cagniard De Latour's discovery in 1836-37, that yeast consisted of living cells which propagated during fermentation, although J. v. Liebig and other famous chemists maintained for a long time the older concept that the yeast acted only as an inanimate substance. In addition to this came the general introduction of the thermometer and the saccharometer, and after 1845, K. J. N. Balling's entire fermentation chemistry.

Nevertheless the beer fermentation process itself caused many difficulties. Sometimes the taste of the beer was so abnormal that quite a number of brews had to be thrown into the sewer. At Carlsberg, too, these difficulties occurred, especially during the years after 1859, and considerable quantities of beer had to be discarded. Brewers exchanged yeast, but this was no help. All yeasts had become unreliable, because brewing during the summer months was taken up everywhere to fulfill the increasing demand for beer. Jacobsen was much occupied by the problem, but all efforts were in vain, and day did not dawn until L. Pasteur appeared after 1860 with his epoch making research on the mysteries of fermentation.

Jacobsen was quite fascinated by the genius of Pasteur, and he realized that the methods of science would be decisive for the development of his brewery. In 1871, in collaboration with the chemist G. C. Faye, Jacobsen installed a small brewing laboratory. In 1875 he aimed higher, remembering the appeal of H. C. Ørsted's lectures and deeply impressed by the marvellous development of the natural sciences. Now he wanted to create a truly scientific laboratory with two departments, a chemical one and a physiological one. His purpose became obvious when he established the Carlsberg Foundation the following year.

III

An outline of the history of the two departments

by *H. Holter*

As an introduction to the collection of biographies which comprise the next section of this book, a short outline of the history of the Carlsberg Laboratory's two departments will be given.

At the outset, however, it will be necessary to relate very briefly a few relevant details of the history of the brewery itself. Without these it would be difficult for the uninitiated reader to understand some of the references to institutions, persons and places*) which occur in later chapters of this book.

The brewery that J. C. Jacobsen built in 1847 in Valby was named Carlsberg after Jacobsen's son Carl. When his son had grown up, J. C. Jacobsen built in 1870–71 a new brewery close to the old one. The two breweries were called Gamle (old) Carlsberg and Ny (new) Carlsberg (L), the latter being leased to Carl Jacobsen. The steadily increasing production, however, necessitated continual extensions; only ten years afterwards (1882) Ny Carlsberg was therefore incorporated in Gamle Carlsberg and was thereafter called »Mellembryggeriet« (L), while Carl Jacobsen in 1880–83 built himself an entirely new Ny Carlsberg a bit further from the original brewery, but still in its close vicinity (see map).

* The localities mentioned in the following are designated by capital letters referring to the map on the last pages of this book (fig. 106).

The two breweries were independently owned and run by father and son, and they remained independent after the death of J. C. Jacobsen in 1887, because Jacobsen in his will had donated Gamle Carlsberg to the Carlsberg Foundation. In 1902 Ny Carlsberg was also donated to the Carlsberg Foundation by its owner Carl Jacobsen (who, incidentally, stipulated that the income from his brewery should be used for the support not of science, but of art).

Although the two breweries were thus united as regards ownership, they were still for some years run as separate units, until in 1906 they were finally merged under the administration of Carl Jacobsen as director, until his death in 1914.

This history explains why Carlsberg is officially called the »Carlsberg Breweries« to this day, although since 1906 it has actually been only one brewery. It also explains that the original Carlsberg Laboratory (B) was established within the site of Gamle Carlsberg, which was J. C. Jacobsen's domain. Apart from the Carlsberg Laboratory, each of the two breweries had its own laboratory for handling practical brewing problems. Of these, the Ny Carlsberg Laboratory (M) was the oldest, which may seem paradoxical; the explanation is that J. Kjeldahl at the Carlsberg Laboratory during the first ten years of his tenure assisted Gamle Carlsberg in the solution of technical problems, until Gamle Carlsberg also established its own laboratory in 1886 (D).

The map on the last pages shows the relevant parts of the brewery grounds, in which all the localities mentioned in this book are situated and marked with capital letters.

In 1875 when the brewer J. C. Jacobsen realized his plans for establishing a laboratory for scientific research, he was assisted in the practical aspects of this task by Johan Kjeldahl, the chemist at the oldest small brewery laboratory (C) which existed only a few years. When in 1876 the new research laboratory (B) was formally established under the direction of the

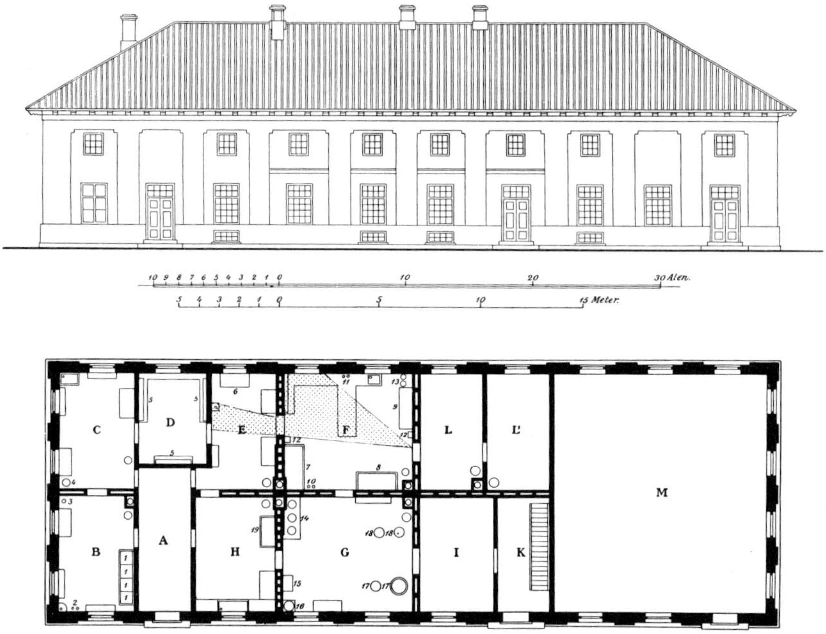

2 Front view and plan of the first Carlsberg Laboratory
B&C: Physiology Laboratories. D: stock room. E: balance room. F, G & H: Chemistry Laboratories. I: coalshed.

newly created Carlsberg Foundation, it was thus natural that Kjeldahl was appointed head of the Chemical Department of the Carlsberg Laboratory. He held this position until his death in 1900. The other department of the laboratory, the Physiological Department, had as its first head Rasmus Pedersen. However, in 1877 Pedersen left this position, and after a short period of vacancy he was succeeded by Emil Chr. Hansen who was head of the Physiological Department from 1879 until his death in 1909.

Originally the laboratory had no assistants. As early as 1877, however, one assistant post for each department was established. During the first decade these assistants changed rather frequently. After a year or two at the laboratory, most of them took positions in the brewery and ceased to be occupied with pure science. The first notable exception to this rule was W.

Johannsen, assistant at the Chemical Department 1881–87, who later became professor of plant physiology at the University of Copenhagen, and the »Founding Father of Danish Genetics«. The other exception was Fr. Weis, assistant at the Chemical Department during Kjeldahl's last years. He became recognised as a bacteriologist, and professor of plant physiology at the Veterinary and Agricultural University of Copenhagen. At the Physiological Department the only assistant who stayed for any length of time during this first period was J. Chr. Holm who was Hansen's assistant from 1884 to 1891. He later became a sub-director at Alfr. Jørgensen's Laboratory for Fermentation, Ltd., Copenhagen.

The main external event that happened to the laboratory during the time of Kjeldahl and Emil Chr. Hansen was the erection of its new and permanent home. The original laboratory had been a rather modest building in the grounds of the brewery (B), but around 1885 J. C. Jacobsen, pleased with the success of his venture, decided to build a new and much bigger laboratory at Gamle Carlsbergvej, outside the brewery but close to it. (F, G, H). He participated with great interest in the early stages of the planning but died (1887) before construction was actually begun. The new building was finished in 1897. It proved to be an adequate framework for the activities of the laboratory, and it remained, apart from certain interior rearrangements, essentially unchanged to the present time. Even now it forms the nucleus of the assembly of research laboratories which the Brewery has established during 1974–76.

Throughout the first two decades of the laboratory's history and before the move to the new building, the work performed in both departments was closely related to the scientific basis of beer brewing. This was strictly in accordance with J. C. Jacobsen's intentions, and it may be assumed that he himself, as long as he lived, saw to it that this close relation was maintained. At least we know that he was indeed most interested in what went

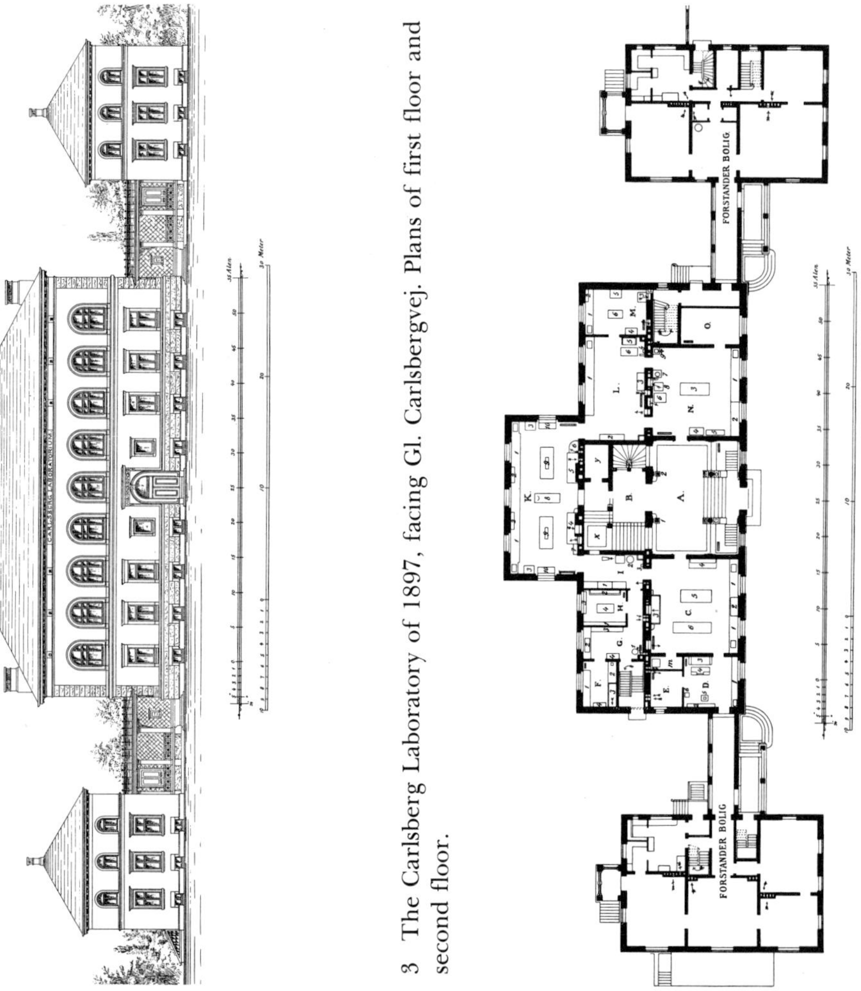

3 The Carlsberg Laboratory of 1897, facing Gl. Carlsbergvej. Plans of first floor and second floor.

on in the laboratory. The intimacy between the laboratory and the brewery is perhaps best illustrated by the fact that during the first years every newly appointed assistant had to undergo a three months' training course at the brewery, before he took up his duties at the laboratory, in order to become familiar with the practical background of his future research work.

First floor (Chemical department): A: Entrance Hall. C & D: professor's laboratory and office. E: balance room. F: fume cupboards. G: wash room. H: stock room. K–N: chemistry laboratories. O: stock room. Forstander-bolig: professor's premises.

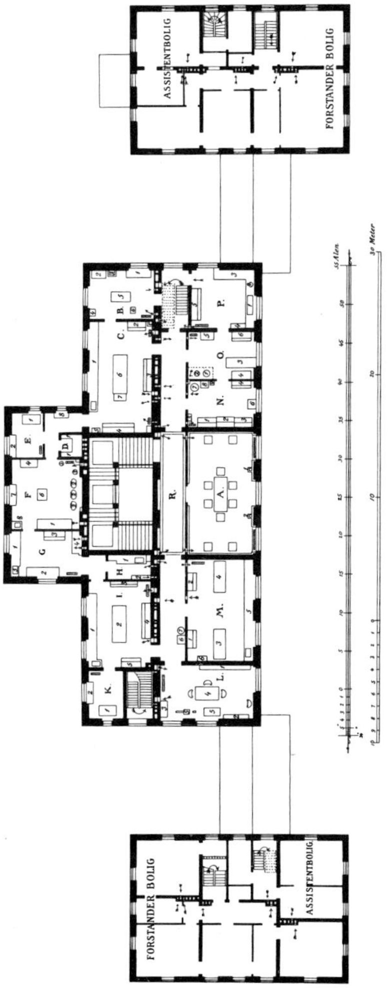

Second floor (Physiological department): A: meeting room. B & C: professor's office and laboratory. D: constant temperature room. E & K: sterile rooms. F, G, I & M: laboratories. H: darkroom. L: library. N: stock room. O: sterilization. P: wash room.

However, it should be clearly understood that even in this early period the work of the laboratory was not primarily concerned with the development of industrially useful knowledge and techniques. The primary research object was basic science. For technical problems, the breweries had their own laboratories.

4 Changing population density. I: Chemical department 1897. Left to right: H. Jessen-Hansen, J. Kjeldahl, N. Hjelte Claussen.

In describing the initial stages of the laboratory's development, I have found it natural to treat the laboratory as a whole. But from now on, that is to say roughly from the time of the move to the new building (G), it would become too complicated to continue in this way, and the two departments will therefore be considered separately.

For the sake of clarity the subject matter has been arranged in sections, named after the heads of the two main departments in chronological order. At the end of each section, a list of the department's assistants during the period in question will be given. For practical reasons only those could be included who have contributed significantly to the department's scientific activities. Short-term visiting scientists had to be omitted, and likewise assistants who had stayed only for a short time (less than three years). For each person is noted: Time spent at the department, type of work performed, main points of later career, and years of birth and death for those no longer alive at the time of writing.

In cases of overlap assistants are listed under the name of the head of department corresponding to the main period of activity.

THE CHEMICAL DEPARTMENT.
S. P. L. SØRENSEN

Kjeldahl's successor as head of the department was S. P. L. Sørensen, who held this position from 1901 to 1938. His great achievements in biochemistry will be evaluated in his biography. Here it will only be noted that it was Sørensen who established research in proteins and proteolytic enzymes as the main topic of the department's activities which – in spite of occasional excursions into other fields – it has remained to this day. This stress on basic research also brought about a gradual loosening of the direct and everyday ties between the laboratory and the brewery.

This can, for instance, be seen by considering the careers of the department's assistants after leaving the laboratory. With the exception of W. Johannsen and Fr. Weis, Kjeldahl's assistants all ended up in the Carlsberg Breweries or otherwise in the brewing industry. Only one of Sørensen's assistants chose the same career. The others either stayed in the department until they retired, or moved on to scientific positions elsewhere in Denmark.

Sørensen, however, contributed to the brewery in other ways which will be discussed in chapter 10. He also participated in regular meetings with the staff of the brewery laboratory and the brewery technicians, and welcomed people from the brewery in his department, for technical discussions and informally at the daily lunch table. This form of contact was much appreciated by both sides, and it continued under Sørensen's successors.

During Sørensen's time a slow but steady expansion of the department took place: The number of assistants grew, and so did the number of visitors. This necessitated an increase in working space which was brought about by incorporating into the laboratory (around 1930) a number of basement rooms that originally served as living quarters for the stewards. In this

5 Changing population density. II: Same view as fig. 4, 1909. Left to right: H. Jessen-Hansen, M. Høyrup, Chr. Pedersen (lab. steward), S. Palitzsch, S. P. L. Sørensen, E. Jürgensen.

way a whole floor was added to the laboratory space. At the same time the laboratory library was moved to the top floor of the building, and an elevator was installed.

Early in Sørensen's days, in 1904, a *Sub-department* was added to the Chemical Department, but the affiliation was only temporary and rather loose. It came about as follows: The brewery laboratory at the Ny Carlsberg Brewery (K on map) had since 1890 employed the chemist *H. Schjerning,* who did good work on the problem of beer turbidity, which he recognized to be caused by a protein component of barley. In 1904 the Gamle and Ny Carlsberg Breweries were merged, and the Ny Carlsberg laboratory was discontinued. In order to give Schjerning an opportunity to continue his research project on barley proteins, he was appointed on May 1, 1904 as »Director extraordinary«, attached to the Chemical Department of the Carlsberg Laboratory. One may suppose that Sørensen did not want his scarce working space encroached upon – at any rate it was decided that there was no space available in the buildings of the Carlsberg Laboratory proper, and therefore the brewery assigned to Schjerning some of the rooms in the old Carlsberg laboratory building (B on map) in the grounds of the brewery. This arrangement worked satisfactorily until Schjerning's death in 1914, when the position, which was purely personal, was allowed to lapse. Schjerning's biography will be found in the collection of biographies of heads of departments.

Sørensen's assistants:

1) H. Jessen-Hansen, 1892–1930.
Carbohydrate chemistry; baking properties of wheat flour; sugar analysis; editor of the laboratory's journal Comptes Rendus des Travaux du Laboratoire Carlsberg.
Was appointed by Kjeldahl, but is listed here because his main work fell in Sørensen's time. Stayed at the Department until his retirement. Born 1860, died 1934.
2) C. E. Pedersen, 1898–1908.
Collaborated with Sørensen in analytical work.
Later became a brewer in Nørre-Sundby in Jutland.
Born 1874, died 1948.
3) A. C. Andersen, 1904–1908.
Collaborated with Sørensen in analytical work.
Later director of Chemical Department, State Research Institute of Animal Husbandry, and for many years a trustee of the Carlsberg Laboratory. Born 1882, died 1970.
4) S. Palitzsch, 1905–1919.
Collaborated as a physical chemist in work on hydrogen ion concentration.
Later lecturer in chemistry at the Technical University, Copenhagen. Born 1882, died 1959.

5) Margrethe Sørensen (née Høyrup), 1909–1937.
Collaborated with her husband in protein investigations. Stayed at the Department until her retirement. Born 1884, died 1954.
6) J. A. Christiansen, 1911–1915.
Measurements of osmotic pressure of protein solutions. Later professor of physical chemistry at the Technical University and the University of Copenhagen. Born 1888, died 1969.
7) C. Olsen, 1919–1944.
Data in separate biography.
8) K. Linderstrøm-Lang, 1919–1938.
Data in separate biography.
9) G. Haugaard, 1925–1943.
Analytical work; studies on wheat gliadin; glass electrode. Editor of C. R. Trav. Lab. Carlsberg.
Emigrated to USA. Worked in National Dairy Research Laboratories, Oak Dale, N. Y. Elaborated an optical method for estimating fat and proteins in milk. Born 1894, died 1969.

K. Linderstrøm-Lang.

When S. P. L. Sørensen retired in 1938, he was succeeded by K. U. Linderstrøm-Lang, who had been Sørensen's chief assistant for a long time and who during the last years of Sørensen's tenure had in fact already directed a considerable part of the department's work. The transition was therefore very smooth and the main line of research was never interrupted.

However, a serious hindrance for the work of the department was caused by the war which broke out shortly after Linderstrøm-Lang had taken over. Although the German occupation of Denmark (1940–1945) did not affect the laboratory directly, it meant a great deal indirectly. Scientific links with Germany had never been strong, and the occupation the-

6 Changing population density. III: Same view as figs. 4 and 5, 1958. Linderstrøm-Lang (far right), assistants and guests. List of names, see chapter 14.

refore meant practically complete scientific isolation. Moreover, during the later part of the war both Linderstrøm-Lang himself and a good part of his staff devoted themselves to resistance activities which took up a considerable part of their time.

Even before the war, and during the last years of Sørensen's tenure, a steadily increasing number of foreign visitors had come to the department. After the period of isolation was over, this influx from abroad was resumed, and it constituted an important and inspiring element in the life of the laboratory.

During Linderstrøm-Lang's tenure, and thanks to his own initiative, a sub-department was once again added to the Chemical Department. This was the *Sub-Department of Cytochemistry*, established January 1, 1943. Its head was H. Holter, who had been an assistant at the Chemical Department since 1937, and in collaboration with Linderstrøm-Lang had worked on histochemical and cytochemical problems. The new sub-department was at first housed in the same rooms of the Chemical Department that Holter had previously used. But in 1946, when the sub-department of Fermentation Physiology was discontinued (see later in the section on the Physiological Department), Cytochemistry moved to its premises (J) at Gl. Carlsbergvej Nr. 8. It remained there until it was discontinued in 1956, when Holter was appointed head of the main laboratory's Physiological Department after the retirement of Ø. Winge.

A feature of considerable importance for both the Chemical Department and the Cytochemical sub-department was the fact that during Linderstrøm-Lang's tenure, for the first time in the history of the department, financial support did not come exclusively from the Carlsberg Foundation. This novelty was due to the Rockefeller Foundation which developed a great and lasting interest in Linderstrøm-Lang's (and later also Holter's) work, resulting in a series of grants which began in 1937 and, apart from interruption by the war were still running at the time of Linderstrøm-Lang's death in 1959. These grants made possible certain valuable extensions, with regard to both staff and apparatus. The »Nordisk Insulin Foundation« also contributed to the department's work, but to a lesser extent. Another very important source of support was the European Recovery Program (1948–52), popularly called »Marshall Aid« which provided many instruments that would have been too expensive for the laboratory's normal budget.

Another feature of the activities of the Carlsberg Laboratory which began in Linderstrøm-Lang's time was the »Carlsberg Colloquia«, a series of seminars given at weekly or fortnightly intervals by staff members or visitors. These were actually started in the early 1930s, when Sørensen was the head of the department, but the initiative was Linderstrøm-Lang's. The topics discussed were of course mainly biochemical, but the colloquia frequently branched out into adjacent fields of science. Many prominent scientists passing through Copenhagen contributed to these colloquia, and I think it can safely be said that over a number of years they played a not inconsiderable role in the development of Danish Biochemistry. The colloquia continued after Linderstrøm-Lang's death, but with decreased frequency, partially owing to the competition from similar series of seminars that had been set up by other institutions.

Linderstrøm-Lang's assistants:

1) C. Olsen, 1919–44,
   appointed by Sørensen, see separate biography.
2) H. Holter, 1937–43,
   data in separate biography. Editor of C. R. Trav. Lab. Carlsberg.
3) C. F. Jacobsen, 1939–50 (1939–43 supported by outside grants). Studies on protein chemistry; pH-stat.
   Now Assistant Director at the Atomic Energy Commission Research Establishment Risø.
4) G. Johansen, 1944–62.
   Problems of physical chemistry.
   Later assistant in Clinical Laboratory, Glostrup State Hospital.
5) M. Ottesen, 1946–59 (until 1947 supported by Rockefeller Foundation).
   Data in separate bibliography.
6) K. Max Møller, 1949–54 (until 1950 supported by Rockefeller Foundation).
   Studies on enzymes and histochemical techniques.
   Now head of division at the Biological Institute of the Carlsberg Foundation.
7) Aase Hvidt (née Hesselberg Jensen), 1951–56 (supported by Rockefeller Foundation and other grants).

Protein studies.
Now lecturer in chemistry, University of Copenhagen.

8) W. Andersen, 1952–56 (until 1954 supported by other grants).
Insulin studies.
Now head of Patent Department, Novo Industri A/S, Copenhagen.

9) Bodil Meedom (née Dahlerup-Petersen), 1952–59 (supported by Rockefeller Foundation).
Studies on subtilisin and chymotrypsin.
Now retired.

10) S. O. Nielsen, 1956–60.
Physico-chemical aspects of protein studies. Now Research Associate in chemistry, Atomic Energy Commission Research Establishment Risø.

Holter's assistants at the Sub-department of Cytochemistry:

1) E. Zeuthen, 1941–46 (until 1943 at the Chemical Department; supported by outside grants).
Development of Cartesian diver technique; cell physiological and embryological studies.
Now director of the Biological Institute of the Carlsberg Foundation.

2) N. Andresen, 1943–52 (until 1946 supported by Rockefeller Foundation).
Studies on cytology and physiology of amoebae.
Now Educational Consultant, State School for Advanced Education, Copenhagen.

3) C. Chapman-Andresen, 1948–56 (supported by Rockefeller Foundation).
Studies on amoeba physiology and pinocytosis.
Moved with Holter to Physiological Department in 1956.

4) S. Løvtrup, 1946–54 (until 1952 supported by Rockefeller Foundation).

Amoeba cytochemistry; Cartesian diver development; embryology of frog egg.
Now professor of zoophysiology at the University of Umeå, Sweden.

5) K. Max Møller, 1954–56.
Studies on amoeba cytochemistry.
Moved with Holter to Physiological Department in 1956.

M. Ottesen

After Linderstrøm-Lang's death in 1959 he was succeeded by his chief assistant, M. Ottesen, who holds the position as head of the Chemical Department to the present day. As was the case at S. P. L. Sørensen's retirement, the succession did not cause swift changes in the orientation of the department's work or organization. However, after 1971 it fell to Ottesen's lot to participate in the profound reorganization of the whole Carlsberg Laboratory which is the subject of chapters 11 and 12.

During Ottesen's tenure the Chemical Department, besides its assistants and visiting scientists, was host to an independent investigator who, without actually forming a sub-department, was independent in the sense that he had his own budget and was responsible directly to the Carlsberg Foundation. This was Mogens Westergaard, who from 1949 to 1962 had been professor of Genetics at the University of Copenhagen. After a coronary thrombosis he had to give up the strenuous work at the University, and a grant from the Carlsberg Foundation permitted him to return to the Carlsberg Laboratory, where he had worked as a student under professor Ø. Winge. Westergaard stayed at the Chemical Department from 1962 till his death in 1975, and he felt that he did some of his best work in those years.

Ottesen's assistants:
(up to the reorganization of the laboratory, 1972–75)

1) G. Johansen, 1944–1962.
   See list of Linderstrøm-Lang's assistants.
2) A. Johansen, 1954–1976.
   Ultracentrifuge studies.
3) B. Foltmann, 1959–1963.
   Studies on rennin and prorennin.
   Now professor of biochemical genetics at Copenhagen University.
4) Aase Hvidt (née Hesselberg Jensen), 1960–1964.
   Deuterium exchange in proteins.
   Now lecturer in chemistry at Copenhagen University.
5) P. Schack, 1963–1966.
   Proteolytic enzymes in Papaya latex.
   Now lecturer at Institute of Physical Chemistry, Technical University, Copenhagen.
6) L. Willumsen, 1964–1970.
   Deuterium exchange in proteins.
   Now engaged in hospital career.
7) I. Svendsen, 1964–
   Studies on subtilisin.
8) J. Johansen, 1966–
   Studies on metallo-enzymes.
9) Kirsten Biedermann (née Egund), 1969–
   Studies on Deuterium exchange and proteolytic reactions.

THE PHYSIOLOGICAL DEPARTMENT.
EMIL CHR HANSEN

After the move to the new building in 1897 Emil Chr. Hansen continued his work on yeast taxonomy and physiology. In this he was helped by his two assistants, A. Klöcker and H. Schiönning who both stayed with him until his death in 1909. Schiönning then left the Carlsberg Laboratory and became an official in the beer taxation service. Klöcker stayed and was given a special assignment.

As we shall see later, a radical change in the department's research orientation was introduced by Hansen's successor, Johannes Schmidt. However, there was still a good deal of unfinished work from Hansen's time to clear up, and Klöcker, having been Hansen's assistant since 1892, was the right man for it. He was therefore made »Director extraordinary« at the Physiological Department (as Schjerning was at the Chemical) for the purpose of winding up the yeast work. He worked at first in his usual room in the main laboratory building, but in 1914, when Schjerning died, the brewery offered him the vacated premises in the old laboratory (B on map), and there he continued his work until his death in 1923. As in Schjerning's case his special position was allowed to lapse after him.

JOHANNES SCHMIDT.

The successor of Emil Chr. Hansen in the main Physiological Department was Johannes Schmidt who was appointed in 1910. Schmidt's cast of mind was very different from that of his predecessor. Hansen had throughout his life studied one group of organisms, the yeasts. Schmidt was interested and productive in many fields: botany, bacteriology, zoology, genetics, and oceanography. He did important work in all of them, but his lasting fame was mainly due to his achievements in marine biology and oceanography, which occupied him more and more during the later part of his life. This work necessitated a series of expeditions and other extramural activities. Needless to say, this kind of work also demanded a big staff (which could only partly be provided by the Carlsberg Laboratory) and considerable support by grants from various outside sources, although a great part of the financial cost was borne by the Carlsberg Foundation.

Although Schmidt was eminently successful and his work was acknowledged by everyone, it cannot be denied that in the course of time he had strayed rather far from the type of research that J. C. Jacobsen had had in mind when establishing the Carlsberg Laboratory. During his first years in the department, Schmidt had followed the traditions of the laboratory, had (in 1911) established a greenhouse in the laboratory's garden (E on the map) and had carried out an extensive study of hop improvement by suitable breeding methods. Later, however, he was completely taken up by his great interest in oceanography. He died in 1933, soon after the completion of his last expedition around the world.

Johs. Schmidt's assistants:

1) K. V. T. Topp, 1910–1915.
   Studies of hops and hop chemistry.
   Later military pharmacist in Java. Born 1882, died 1960.
2) Ø. Winge, 1911–1921.
   Data in separate biography.
3) Chr. Folker, 1915–1923.
   Studies of hops and hop chemistry.
   Later in private business, Copenhagen. Born 1887, died 1932.
4) V. Ege, 1921–1933.
   Zoological and marine biological studies.
   Later at the Marine Biological Laboratory*, Copenhagen. Born 1887, died 1962.
5) Aa. Strubberg, 1923–1933. (supported by outside grants).
   Studies on fishery and oceanography.
   Later Principal in the Danish Ministry of Fisheries. Born 1881, died 1969.

6) E. C. Svarrer, 1923–1930.
Studies of hops and hop chemistry.
Later head dispenser at a pharmacy in Copenhagen. Born 1899, died 1967.
8) A. F. Bruun, 1927–1933. (partially supported by outside grants). Zoological and marine biological studies.
Later lecturer in zoology at Copenhagen University, Leader of the Danish »Galathea« expedition round the world. Born 1901, died 1961.
9) P. C. Jespersen, 1930–1933.
Marine biological studies.
Later director of Plankton Laboratory, Danish Fishery Institute. Born 1891, died 1951.
7) Esther Hansen, 1923–1933.
Zoological and marine biological studies.
Later at the Marine Biological Laboratory*, Copenhagen.

During Johs. Schmidt's time there occurred two extensions of the Physiological Department's working space. One was necessitated by the department's increasing and many-sided activities, and was effected by a rearrangement of the rooms available in the main building of the laboratory, in parallel with the extensions of the Chemical Department under Sørensen. Thus several rooms in the top floor of the building were added to the premises of the department.

The other extension was initiated by the Board of Trustees of the laboratory, who felt that the department's activities had deviated from the intentions of J. C. Jacobsen. In order to compensate for this, a new sub-department was established in 1931, the *Sub-department of Fermentation Physiology*. Its head was the microbiologist Niels Nielsen, and it was housed in a two-sto-

* The Marine Biological Laboratory mentioned here was established in 1933 by the Carlsberg Foundation and the Danish Fishery Institute, in order to provide facilities for the continuation of Schmidt's work and room for the collections from Schmidt's expeditions.

7 The building Gl. Carlsbergvej Nr. 8
(Sub-departments of Fermentation Physiology and Cytochemistry).

rey villa (J) at Gl. Carlsbergvej Nr. 8, adjacent to the main laboratory building. The ground floor and basement rooms were converted into laboratories, while the top floor provided living quarters for the janitors of the main laboratory.

The acquisition of this building constituted the first external extension of the Carlsberg Laboratory since the erection of the main building in 1897. The building housed the Sub-department of Fermentation Physiology, until it was abolished in 1946, when the premises were taken over by the Sub-department of Cytochemistry as already mentioned. When this also was discontinued in 1956, the laboratory part of the building was taken over by the brewery.

Nielsen's assistant throughout the existence of the sub-department was Vagn Hartelius, who worked on problems of yeast physiology, and in 1946 joined the staff of the Physiological Department in the main building.

ØJVIND WINGE.

Let us resume the story of the main Physiological Department. When Johs. Schmidt died in 1933, he was succeeded by the mycologist and geneticist Øjvind Winge, who already knew the department from the time when he had been Schmidt's assistant from 1911 to 1921. With Winge in office, the main field of the department's activities became once more the study of yeast, even though for a number of years he continued and completed the genetical studies of other organisms which he had begun before returning to the Carlsberg Laboratory. The yeast research was his pet project, and he made it a great success. Among his many contributions to yeast biology were the demonstration of the sexual reproduction of yeasts and the artificial hybridization of yeast species. The latter achievement involved the development of special techniques of micromanipulation which attracted many students who were eager to learn Winge's methods.

Winge also resumed close collaboration with the brewery. He conducted systematic breeding studies with barley and hops in an experimental farm acquired by the brewery, and this farm therefore became, in Winge's time, in a sense an extension of the department.

Winge stayed in office until his retirement in 1956 but he continued to supervise the brewery's experimental farm until his death in 1964.

Winge's assistants:

1) J. H. Wanscher, 1933–36.
Yeast genetics.
Now lecturer at Department of Horticulture, The Veterinary and Agricultural University, Copenhagen.

2) O. Laustsen, 1933–39.
Studies on yeast genetics and hops.
Later head of department at the Research Laboratory, Carlsberg Breweries.
3) E. Ditlevsen, 1933–56.
Genetical studies; breeding of barley and hops.
Later assistant to professor Winge in Experimental Farm Project. Born 1902, died 1969.
4) Aase Hjort, 1933-62.
Yeast taxonomy.
Retired 1962. Born 1896, died 1963.
5) Aa. Skovsted, 1936–56.
Yeast genetics.
Later head of Department at the Botanical Museum, Copenhagen.
6) Sv. Stangegaard Nielsen, 1939–44.
Studies on hops and barley.
Later with the pharmaceutical firm Ferrosan Ltd., Copenhagen. Born 1913, died 1970.
7) E. Bille Hansen, 1944–49.
Studies on hops and barley.
Later lecturer at the Institute of Plant Physiology, Copenhagen University.
8) V. Hartelius, 1946–56 (from Sub-department of Fermentation Physiology).
Yeast studies.
Later assistant at the Biological Institute of the Carlsberg Foundation. Retired 1974.
9) Catherine Roberts, 1949–61.
Yeast genetics.
Retired 1961.

HEINZ HOLTER.

At the Physiological Department Winge was succeeded by H. Holter who, as mentioned before, had been the head of the Sub-department of Cytochemistry since 1943. During Holter's tenure, the Physiological Department continued to work on microorganisms, including yeast, but with a tendency to change the main field of the department's activity in the direction of general cell biology. However, collaboration with the brewery was never allowed to lapse, and Holter regularly participated, together with Ottesen, in the meetings with the brewery staff that had been instigated by S. P. L. Sørensen.

In the 1930s, at the time of the »internal expansion« of the Carlsberg Laboratory, the modernization of the Physiological Department had not been as thorough as that of the Chemical Department. This was rectified when Holter's succession gave a suitable opportunity, and some rooms and modern facilities were added to the department, but without structural changes of the building. With regard to personnel and apparatus, the situation of the department was much improved by a series of USA grants from the National Institute of Health which played a role similar to the Rockefeller grants of an earlier period.

Holter retired in 1971. The subsequent history of the department – indeed of both departments – forms a part of the great reorganization of the whole structure of the Carlsberg Laboratory which is the subject of chapters 11 and 12.

Holter's assistants:
(at the Physiological Department)

1) K. Max Møller, 1956–72 (from Sub-department of Cytochemistry).

Studies on cell physiology and cytochemistry.
Now head of division at the Biological Institute of the Carlsberg Foundation.

2) C. Chapman-Andresen, 1956–72 (from Sub-department of Cytochemistry).
Studies of amoeba physiology and pinocytosis.
Now research associate, Institute of General Zoology, Copenhagen University.

3) V. Hartelius, 1956–72.
Yeast studies.
Later assistant at the Biological Institute of the Carlsberg Foundation.

4) P. Ottolenghi, 1957–72 (supported by outside grants).
Studies on cell physiology and biochemistry.
Now lecturer in physiology, Aarhus University.

5) J. Friis, 1956–72.
Yeast studies.
Now lecturer in molecular biology, Odense University.

This brings the story of the two departments more or less up to date, but before concluding something must be said about certain features of the Carlsberg Laboratory which are common to both departments and which have not been mentioned so far.

In 1966 an important addition of the laboratory's research facilities was established. This is a building specially designed for work with radioactive isotopes, planned by the two departments in collaboration and destined for common use. The building (see fig. 73) is situated in the laboratory grounds in the vicinity of the greenhouse and the tennis court but isolated from the main building. All the facilities of the isotope laboratory were also available for the staff of the brewery research laboratory.

One of the greatest assets of the Carlsberg Laboratory is its library, one of the best minor scientific libraries of Denmark. It was originally housed in a single large room on the main floor of the Physiological Department, but it soon outgrew this accommodation and was transferred to the top floor of the building, where it continued to grow and to present problems of space and organization. In spite of such minor inconveniences, the Carlsberg Laboratory is justifiably proud of its library. It has enabled the staff of the laboratory throughout the years to obtain answers to many professional questions at once, without the delay caused by trips to one or another of the big central libraries of Copenhagen.

Another important feature of the Carlsberg Laboratory, and one of a rather special nature, is the fact that the laboratory has its own journal, the »Comptes Rendus des Travaux du Laboratoire Carlsberg«. This is treated in chapter 7.

IV

Chemical department. Biographies.

Johan Kjeldahl

1849–1900

by *W. Johannsen**

The Council of the Carlsberg Laboratory has asked me to write a biography of the Head of the Chemical Department of the Carlsberg Laboratory, Johan Kjeldahl who passed away on July 18, 1900. I will, therefore, attempt to provide a review of Kjeldahl's scientific work as the first task, and subsequently in connection with some biographical data try to describe a few features of his sympathetic and exceptional personality.

I.

Kjeldahl's name is renowned all over the world, owing to his method for nitrogen determination in organic material, published in 1883. This method is used in all laboratories where plant and animal material is subjected to chemical analysis. The usefulness of the method has been and still is extraordinarily great. Whoever has had first hand experience with H. Will & F. Varrentrapp's method will recall with gratitude the enormous relief which Kjeldahl's method meant, especially when dealing with analyses of liquid material. The fact that nitrogen determination became literally the most manageable

8 Johan Kjeldahl in the old Carlsberg Laboratory. Painted 1896–97.

**Written in 1900.*

quantitative determination has, of course, greatly benefited the study of nitrogenous compounds and their metabolism. Kjeldahl has been proven right in his expectations of the method and its possibilities in »working out problems which hitherto could not (or only to an unsatisfactory degree) be treated, due to the inconveniences and the slowness of the analytical methods, features which were of additional hindrance when dealing with generally labile material«. Hence it is primarily the physiological sciences which have reaped the benefit of the method; but it can safely be stated that the method has become invaluable also for agricultural-technical investigations and innumerable standardization tests.

Kjeldahl felt compelled to design such a method because he had initiated studies of protein metabolism during barley germination and during alcohol fermentation. He shelved these investigations in order first to create a tool. The starting point for these efforts were J.A. Wanklyn's experiments (and those by other English researchers) in which organic materials were treated with potassium hydroxide and permanganate. These experiments showed that a certain, albeit fluctuating percentage of protein nitrogen is released as ammonia, whereas other constituents, such as asparagine, amides and amines release their entire amount. After a number of futile attempts to improve this method, Kjeldahl became captivated by »the obvious idea that the results would turn out better if one performed the oxidation in acid, since the inclination to ammonia formation under such circumstances might be considered to be much greater«. The application of dilute acid instead of potassium hydroxide gave better results immediately. But not until concentrated sulphuric acid was used, combined with heating the material close to the boiling point and subsequent addition of dry permanganate, was the desired result obtained. The idea, which seems so obvious, perhaps like all creative ideas, came as a flash. Nevertheless, plenty of hard work and toil was spent in the practical design and standardization

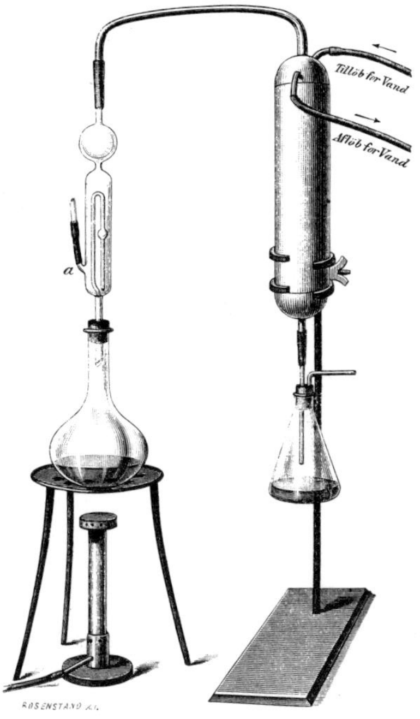

9 The original Kjeldahl apparatus for the determination of nitrogen: Distillation of ammonia formed in the sulphuric acid digest. Later modified from glass to copper flask, as seen in fig. 8.

of the method. Kjeldahl also spent much time on the adaptation of the iodometric titration for the determination of the ammonia formed during the decomposition process (a side of the method which is a separate matter and has not found general use).

The method in its original form marked an enormous stride. It aroused a lively discussion, especially in German journals, on the precision of the method, and this, combined with experience gathered at the Carlsberg Laboratory, led to minor improvements with respect to distillation and titration. It should be mentioned that German chemists especially should be credited for further developments of the method, as for instance the addition of cupric dioxide to the sulphuric acid, which speeds up and intensifies the effect of the latter on the organic material, thus making the presence of permanganate superfluous most of the time. Moreover, the adaptation of the method for use in the presence of nitrates or nitro compounds is an important addition to the value of the method and was developed outside the Carlsberg Laboratory.

Beside the fact that the Kjeldahl method is a superlative tool for physiological-chemical studies in general, the circumstance that the method, even in its original form, was practicable for proteins had the immediate application of providing proof that proteins do not contain azo groups nor nitro or nitroso groups as had been suggested by others.

The nitrogen method did not come to play as great a role in Kjeldahl's own continued research as he had anticipated. He did not resume his studies on proteins and their metabolism, although he continued to contemplate such investigations and occasionally did certain preliminary studies, especially on the occurrence and action of proteolytic enzymes.

During the first years of his activity at Carlsberg, Kjeldahl was engaged in determinations of extract, alcohol and other components of beer and wort, mainly for the purpose of technical standardization, but he soon entered into areas and problems of more scientific interest, namely his elegant investigations of sugar-forming enzymes. His publications on this subject are recognized as pioneer work in the sense that before the appearance of these studies, to quote Kjeldahl, »investigations of the conditions which influence the effect of enzymes had been rather meager and sporadic«. Kjeldahl treats relevant questions, especially the influence of temperature and certain chemical reagents on enzyme action, in a systematic and exact way. In this way he obtained results of great importance for the understanding of the general rules governing the dependence of enzyme action on exogenous factors. Also his derivations concerning assessment of enzyme potentiality have contributed significantly to a clarification of the relevant problems. These investigations are greatly respected as is evident from the way they are quoted in all major reviews on enzymes, for instance in E. Duclaux's new handbook, and especially in E. Bourquelot's small essay (with a thorough and careful reference list), »Les ferments solubles« (1896). In contrast to this, it is noteworthy that C. Oppenheimer's new monograph »Die Fermente und ihre Wirkungen« (1900) with an overwhelming bibliography only allocates meager room for Kjeldahl's investigations, moreover it contains positive errors which could have been avoided by a real knowledge of Kjeldahl's investigations.

In the first of the two mentioned enzyme articles Kjeldahl

also described an improvement of C. G. Reischauer's modification of H.C.v. Fehling's sugar titration. Later on he tended to follow the more exact gravimetric determination, and also here provided an essential contribution.

Through the study of sugar-forming enzymes Kjeldahl was led into physiological problems concerning carbohydrate metabolism during germination, and also purely chemical problems such as the quantitative determination of carbohydrates in mixtures etc. In 1881 there appeared the simultaneous publications »Studies on invertin« and »Investigations of carbohydrates in barley and malt with special reference to the occurrence of cane sugar«.

This work contains the demonstration, although not isolation, of a new polysaccharide in barley (»amylan«, isolated at about the same time by C. O'Sullivan). It is also shown that cane sugar is formed abundantly during the sprouting of barley and subsequently undergoes fission to glucose and fructose. Moreover, for analytical purposes it is demonstrated how inversion readily ensues when parts of the plants are ground and extracted in water (as was often done a couple of decades ago). It is emphasized that a valid analysis of the content of the readily hydrolyzable plant material can only be obtained if one brings about an arrest of enzyme activity by addition of agents destroying the enzymes. Another important tool in plant physiological chemistry stemming from Kjeldahl's research and rightly promoted by him is the use of barium carbonate and strong alcohol in the grinding and extraction of plant material. This publication does not seem to be as well known among plant physiologists as would seem desirable.

The above mentioned work also contains preliminary attempts to perform simultaneous determinations of different sugars. Kjeldahl continued to work on this task which also became the basis for his next publication »Investigations of the response of the sugars to alkaline copper solutions«. In this important work Kjeldahl demonstrates a hitherto overlooked source of error in the analytical methods, based on the reduction of the copper solution by the sugar. Notwithstanding the methodological improvements »determinations of sugars performed by various analytical workers in different laboratories show poor reproducibility, yet the two controls turn out to agree closely«. Kjeldahl shows clearly that the fluctuations are due to variablity in the tendency of the solution to take up oxygen during the reduction process. The more readily the oxygen diffuses into the solution, the less precipitation of cuprous oxide. This realization led to a radical improvement of the method, the use of a stream of hydrogen during the boiling of the mixture. This brought about quite a different degree of precision than that obtained by the usual routine method.

It was obvious that a revision of previous work which had involved a calculation of the ratio between precipitated cuprous oxide and sugar equivalents present (normalized for various sugars) was overdue, and Kjeldahl did this job as well. The practical significance of his tables calculated for analysis of pure sugars alone, has been debated. Yet the essence of the matter, the exclusion of oxygen from the sugar-copper mixture, has been abundantly confirmed; thus Kjeldahl deserves special credit for an essential contribution to a development in the field of sugar analysis. The copper reaction itself has also been further investigated by Kjeldahl. He found that the use of A. Soldaini's reagent gives rise to the formation of mesoxalic acid in the oxidation of glucose. A continuation of these recent investigations ought to provide interesting information.

It was with a special interest that Kjeldahl tackled the study of plant proteins, their optical rotation, and their solubility in alcohol solutions of various strength. About these topics he had published only a preliminary communication for a Natural Science meeting in Copenhagen in 1892. As an example we shall give one of the results: All the protein preparations which can be obtained from wheat flour by extraction with alcohol and subsequent repeated precipitation by freezing, showed essentially the same content of nitrogen (17.25%) and of carbon (ca. 52%). Moreover they all, when dissolved in 55% alcohol showed the same optical rotation, $[\alpha]_D = -92°$. The latter figure turned out to be conspicuously constant even in preparations obtained from wheat stemming from highly different latitudes and crops from all 4 years of the investigation. In particular the latter fact seems definitely to speak in favour of the existence of only one alcohol-soluble protein in wheat flour which is in contrast to K.H.L. Ritthausen's view which argued for 3 proteins. Corresponding investigations of the alcohol-soluble proteins from rye gave an optical rotation $[\alpha]_D = -121°$.

By these recent investigations Kjeldahl initiated work of great interest for biological chemistry. They should pave the way for a finer description of peculiar modifications of the more generally occurring plant materials, characterizing the particular species. Kjeldahl himself ascribed great importance to these investigations. Whether during the continued work he encountered special difficulties which have delayed the further course of these investigations, it is difficult to judge; the regrettable fact is, however, that 8 years did elapse without giving rise to any detailed publication.

Besides the above mentioned work, largely developed in a natural context, a couple of smaller separate articles of Kjeldahl were published; one about choline as a normal constituent of beer, one dealing with the application of mercuric oxide in the »liquid version« of carbon determination. The latter article contains also a description of an improved decomposition flask.

Viewed as a whole, Kjeldahl's work bears the indelible stamp of conscious effort to further the development of exact methodology. Correspondingly Kjeldahl attained his great reputation as the creator of excellent research techniques. Kjeldahl's preference for methodology is undoubtedly a consequence of his great interest in physics, whose exact techniques appealed to him. In the younger years his interest was perhaps as much caught by studies of physics as by chemistry. Even during his first years at Carlsberg, he occasionally concerned himself with studies – again methodological ones – about certain aspects of the theory of electricity. During the later years he followed with vigorous interest the developments of modern physical chemistry. A subject which was therefore bound to capture his intense interest was A. Croft Hill's statement that the action of maltase must be a distinctly reciprocal process. The preliminary investigations which he initiated on this occasion gave a negative result which may not necessarily mean that Hill was wrong. On the other hand, Kjeldahl's earlier investigations of the sugar generating enzymes gave him the impression that even a significant accumulation of conversion products would not affect the action of the enzyme in question. And as far as this is concerned, perhaps his investigations could be used as an argument against Hill. How unfortunate that we shall have to miss Kjeldahl's own participation in the discussion of this extraordinarily important matter which supporters as well as opponents of Hill's concept often seem to decide upon a bit too lightly.

II.

Johan Gustav Christoffer Thorsager Kjeldahl was born at Jægerspris in Sealand August 16, 1849, son of the district physician J. P. Kjeldahl and his wife Johanne, née Lohman. He graduated from Roskilde Cathedral School in 1867 and received his Master's Degree in 1873 from the Technical University in Copenhagen with highest honors. He became assistant to C. T. Barfoed the same year at the Chemical Laboratory of the Veterinary and Agricultural University. Through Barfoed, Kjeldahl met J. C. Jacobsen, the founder of the Carlsberg Laboratory, and was employed by him as chemist on May 1, 1875. Shortly after, the laboratory came under the direction of the newly established Carlsberg Foundation, and Kjeldahl, on October 1, 1876, was appointed head of the Chemical Department of the laboratory. His works during the 25 years spent at the Carlsberg Laboratory are described above. It need only be added that it was for him a special pleasure to help the young people who worked as his assistants to continue their education and training, either as technicians or as scientists. He was most liberal in allowing them time to finish their own research and studies, and was always willing to give them good advice. There were others who profited by his insight, such as technicians and scientists not directly connected with the laboratory who came to spend some time there.

In 1890 Kjeldahl was elected to the Royal Danish Academy of Sciences and Letters, in 1892 to the Academy of Science in Christiania (Oslo), and the same year he received his appointment as professor. In 1894 he was given an honorary doctor's degree at the University of Copenhagen, and in 1898 he was made Knight of Dannebrog.

Kjeldahl had suffered from recurrent depressions for a few years which were more like exhaustion than real depressions, and which also seemed much improved following a long rest-cure in Southern Europe and in Norway; but then he suddenly died on July 18, 1900 while swimming at Tisvilde beach. His death was due to a brain thrombosis or hemorrhage. Even when he was younger, he had suffered from psychic depressions which at times interfered with his research. During these periods he felt strongly how much he missed the stimulation and diversity which can be provided by a combination of regular teaching and research; he would certainly have been well qualified for university teaching. However, when there was an opening for such a position a number of years earlier, he did not apply for it, possibly because, in his almost too unselfish way, he did not want to prevent others from getting it. Furthermore, he felt very close to the board of trustees of the Carlsberg Laboratory, and they had at that time specifically asked him to remain at the laboratory. He was often asked to be an external examiner at the Veterinary and Agricultural University and at the Technical University, and he appreciated these tasks.

This, quite briefly, is Kjeldahl's story, at least insofar as public knowledge is concerned. During his years as a student, in the post-graduate period, and even after he had become head of the laboratory, he was a highly esteemed member of a small circle of friends who called themselves »KWIMS«, taken from the first letter of their family names. In this circle of educated intellectuals were several well-known artists; Kjeldahl's love for and understanding of art and literature found much inspiration in this milieu. During his last travel South to regain his health, accompanied by a close »KWIMS« friend and his wife, he took great interest in all the arts of Italy. Throughout his life aesthetics meant a great deal to Kjeldahl. He followed the advice of his favourite Danish poet F. Paludan-Müller »... stick to clever ladies, quick jokes«, and at parties he was most entertaining. He wrote a number of songs for private parties, some funny and baroque, others more profound, and still others where the two moods appear simultaneously. One song, composed to celebrate the 25th anniversary of the Chemistry Laboratories of the University was especially successful and made him extremely popular among the younger people. This article is not the proper place to quote the song, where one encounters a most hilarious mood; it is printed in full in the newspaper »Nationaltidende«, Feb. 18, 1884 in its account of the celebration.

Emotionally Kjeldahl was rather shy, often covering up his feelings with a quick joke which on the one hand would hide his personality and at the same time have the effect he was striving for: to make people laugh. He really was an emotional man, but his emotions fluctuated and would at times manifest themselves in a kind of black humour. Thus he might use funny but rather drastic expressions about himself, especially when he was dissatisfied with his own capacity for work. All of

10 »Kjeldahl in his remarkable and genuine modesty«, supervising the construction of the new laboratory building, some time between 1893 and 1897.

this brings us to another side of Kjeldahl's personality: his remarkable and genuine modesty. He was completely incapable of making himself appear important. This is clearly shown in his absurdly short autobiography written for the official volume issued by the University to celebrate the anniversary of the Reformation in 1894, at which time he received his honorary doctor's degree. Without great demands and in a friendly straight-forward way he managed to have a free and open relationship with his assistants; they trusted him as one trusts an older true friend. Naturally he was well liked by all. His friendly nature and good humour spread joy everywhere he went. He could be depressed at times but he was never unfriendly.

It is easy to understand that Kjeldahl's great zest for life, his abundance of ideas and intuitions, his great interest in art and literature, all tied up a good deal of his energies. This, combined with his failing health, prevented him from working on his scientific studies diligently and without interruption. He had a certain genius, and he was a human rich in abilities and most sensitive. He was incapable of being just a work-horse. He possessed many possibilities other than industriousness. Therefore, the quantity of his work is not large but what he did produce was absolutely first rate. Science has lost an eminent and original intellect who contributed greatly to the honours of his country. His friends are mourning the loss of a most educated and noble personality, loved by all who were fortunate enough to have known him. It may finally also be mentioned that he was indeed a loving son and a faithful brother. In this respect too, Kjeldahl leaves a happy memory.

S. P. L. Sørensen

1868–1939

by *K. Linderstrøm-Lang**

Søren Peter Lauritz Sørensen was born in Havrebjerg near Slagelse in Denmark, the 9th of January 1868, the son of Hans Sørensen, an owner farmer. He early showed interest in bookish occupations and after a few years at school in Slagelse, went to Sorø Akademi, a public school for boys, where he matriculated in 1886. At first he took up medicine but after passing the first examination he changed over to chemistry and took his degree in 1891. He made his expenses while studying partly by coaching and partly by making investigations for the »Geological Survey of Denmark« during the long vacation, and from 1892–1901 he held the position of Assistant at the Chemical Laboratory of the Technical University under Professor S. M. Jørgensen. Simultaneously (1892–1901) he acted as Chemical Consultant at the Royal Naval Dockyard. From 1901 until 1938 he was Head of the Chemical Department at the Carlsberg Laboratory, Copenhagen. He died in Copenhagen the 12th of February, 1939.

With S. P. L. Sørensen's death an epoch in Danish biochemical research was brought to a close. By virtue of his peculiar gifts which enabled him to go to the breadth and the depth in his investigations, he attained such a central position in the consciousness of the scientific world that the concluding of his life's work comes to us as a cleavage. It is difficult to conceive of a continuation of his work along the same lines. The unlimited patience with which every single detail had been investigated and established, making his treatises so massive and ines-

**Written in 1939.*

capable, was no ordinary trait in a science where one often rests content with qualitative reports. Nevertheless he had, and for a long time will continue to have, a great and active influence on biochemical research, not merely through the results he obtained but also through his scientific example, his exhortation to accuracy and consistency.

Sørensen's first works were of a preparative and analytical character, and most of them were in the field in which his teacher, S. M. Jørgensen worked. They all bear the imprint of great thoroughness, industry, and a flair for experiment, but one cannot help observing that they are relatively poor in theoretical considerations. This disinclination to theorize was a characteristic trait of Sørensen's which accompanied him troughout his life.

Sørensen's work on analytical standards was done at the close of the century. »The Employment of Normal Sodium Oxalate in Volumetric Analysis« appeared in 1896–97 and was followed by a number of reports in Zeitschr. für analytische Chemie (until 1908) in which, partly in collaboration with A. C. Andersen, he treated the question of analytical standards very thoroughly. In particular the comparison between sodium oxalate and sodium carbonate as standards was made with exceptional precision and, as is now wellknown, turned out to be in favor of oxalate as had been suggested by Sørensen. With similar accuracy Sørensen in collaboration with H. Bjørn-Andersen and A. Wøhlk carried out an investigation to determine the chlorine and salt in sea water (1902).

His really pioneer investigations however were not begun until after 1901 when he became the Head of the Chemical Department of the Carlsberg Laboratory. His works there fall into four groups: a) the synthesis of amino acids, b) analytical studies, c) investigations on hydrogen ion concentration, and d) protein studies.

The synthesis of amino acids was commenced in 1902. Already then the proteins had caught the interest of Sørensen and he took up the investigation of these substances from the very beginning, studying their structural units, the amino-acids. The starting point was the method evolved by S. Gabriel for synthesizing amines by means of potassium phthalimide. Sørensen coupled sodium phthalimide malonic acid esters with various halogen compounds.

The compounds formed gave, after suitable hydrolysis and decarboxylation, amino acids. This method which was particularly elegant was used for synthesizing a number of amino acids among which ornithine, proline, and arginine should be mentioned. The preparation of the two isomers

α guanido δ aminovaleric acid and

δ guanido α aminovaleric acid

must be regarded as especially important since the identification of the latter with racemic arginine finally settled the constitution of this amino acid.

The analytical investigations went hand in hand with the synthetic ones. During the years 1903–05 there appeared, (in collaboration with C. Pedersen and A. C. Andersen), two careful investigations of Kjeldahl's method of determining nitrogen which had been criticized from various quarters. Although these papers contained a warm defence of Kjeldahl's original procedure, the laboratory turned after a time to J. W. Gunning & C. Arnold's modification because of its distinct advantages in providing for the combustion of nitrogenous compounds (pyridine, piperidine), that are difficult to split.

In 1907 the first paper on formol titration appeared and it was followed in 1907, 1909 and 1910 by several others in which H. Jessen-Hansen and V. Henriques collaborated. As the first really quantitative method of chemically tracing proteolytic cleavage, it is extraordinarily important historically. Nor is its day yet over. It is still used in many laboratories because it is superior in many ways to other methods. This work is typical of Sørensen. It is a model treatise on the H. Schiff reaction, a clear and conclusive demonstration of why

11 S. P. L. Sørensen, 58 years old, painted in his office 25 years after his appointment as head of the Chemical Department.

Schiff was not able to obtain quantitative results that could be utilized, a thorough investigation of the conditions under which this was possible, and a detailed study of the errors of the method. As far as I know, no improvement has been intro-

12 Sørensen at the time of the amino acid syntheses.

duced in the methodology during the course of the thirty years that have passed since it was first used. More than that can hardly be expected.

Sørensen gave his treatise on the formol titration the title »Studies on Enzymes I«, implying that it would be the first of a series of papers. However, when every effort is made to exhaust the subject in a single paper, not many more are likely to follow, and actually only two appeared in this series. On the other hand »Studies on Enzymes II« became a classic work in the real sense of the word and contained material for many papers.

It falls into four main sections.

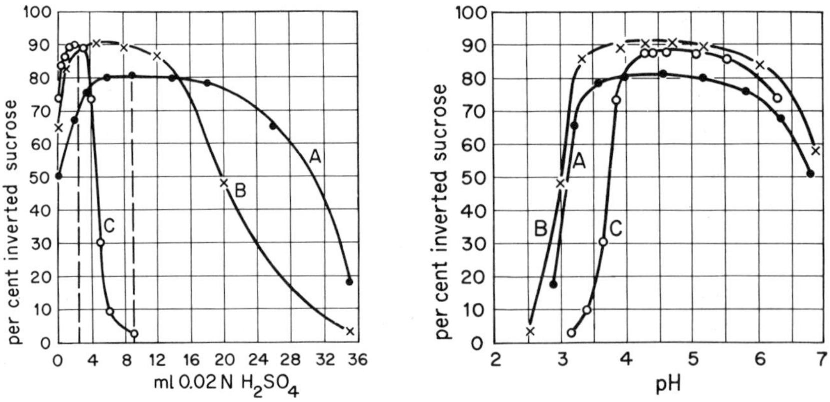

13 Effect of the addition of acid on the activity of invertase.
Left: Acidity indicated the old way by the amounts of acid added.
Right: Acidity indicated according to Sørensen on an exponential scale (pH).

The first deals with the electrometric method of determining hydrogen ion concentration and gives an amazingly exhaustive description of the sources of error. Accurate and exemplary is also the determination of π_0 and of the dissociation constant of water.

The second main section comprises an investigation of the possibility of preparing well defined standard solutions which, on being mixed together, form buffer solutions of fixed pH. These buffer solutions of Sørensen's have won their way into laboratories all over the world where it seems as if they will remain for a long time.

The third main section treats of the colorimetric method of measuring pH. Here, as with the formol method, he started from a known premise and worked it up with such consistency that he eclipsed the earlier workers (H. Friedenthal and E. Salm). He created a method that could be used for the most varied solutions, he made his readers trust him, by himself demonstrating the weaknesses of the method and indicating how they could be overcome.

The fourth main section deals with the application of these

methods when studying the catalysts of the organism, the enzymes. We see here the first pH curves for the enzymes invertase, catalase, and pepsin, the beginnings of a growing literature on the subject, something really fundamental and beautiful. It is difficult for us who are used to descriptions of these facts to imagine enzyme research without this wonderful aid. It is not possible to follow in detail the many paths by which the demonstration of the significance of pH in biological processes has acted as a stimulus to biochemical research in our time, but without hesitation one may say that it turned biochemistry into an exact science. Sørensen certainly did not discover a new natural law, but he gave us the means with which to investigate a known phenomenon, and his simple, clear presentation made the new ideas easily accessible. Through him they became generally recognised.

And then he stopped. The mental background for his studies on Enzymes I and II had actually been to make the methodical tool for the investigation of proteins as complete and exact as possible. And now his interest in these substances awoke again. Apart from a few later papers on bio-catalysts (in conjuncttion with S. Palitzsch, M. Sørensen, K. Linderstrøm-Lang, Lydia Katschioni-Walther, H. Holter, and Bent Andersen) he ceased to take an active part in carrying out the great programme of scientific work initiated in the »Studies on Enzymes«. It is of course wrong to imagine that he turned his back on it. He gave for example a splendid presentation of the question in »Ergebnisse der Physiologie«, 1912. Many people came to the laboratory in order to learn the new methods, but he left it to others to make use of the tools he had forged and when the great flood came that would have drowned a lesser work than »Studies on Enzymes II«, and which nevertheless caused it to be valued a little less than it deserved, he was deep in other things.

The protein studies were about to be undertaken. When one reads the annual reports from the Carlsberg Laboratory one gets there a vivid impression of the great difficulties that had to be combatted. The first paper appeared in 1912 and dealt with the coagulation of proteins on heating (Sørensen and E. Jürgensen). In the report for 1913 it was stated that Professor Sørensen had begun with M. Høyrup a series of investigations of the influence of the hydrogen ion concentration on the chemical and physical properties of proteins. In the next report we read that the protein investigations were continuing and that it had proved necessary to extend them to a still broader basis. In the report for 1915 it was announced that S. Palitzsch and J. A. Christiansen had been drawn into the investigations and that »from these studies, being conducted on an ever wider and wider basis, it appears that an understanding and survey of the whole question will be clearer as the work progresses and for this reason Professor Sørensen thinks it right for the present to postpone publication«. The report for 1916 announced that the investigations are practically finished and the compilation of the material has begun. In 1917 the famous Volume 12 of the »Comptes Rendus des Travaux du Laboratoire Carlsberg« appeared. It contains an amazingly large and trustworthy amount of experimental material and in several respects forms the introduction to modern protein research. The knowledge of the preparative and physical chemistry of egg albumin was vastly enriched. For the first time the conditions under which pure egg albumin can be separated from the other proteins of egg white were exactly defined, the best conditions for crystallization determined, and purification by dialysis investigated.

The electrochemistry of pure egg albumin was investigated i. e. the connection between the hydrogen ion concentration and the hydrogen ion uptake not only in water but at different concentrations of ammonium sulphate. Again it was pioneer work that experimentally has not been bettered. Its theoretical side, however, met with misfortune a year or so later when a fundamental change took place in our conceptions regarding

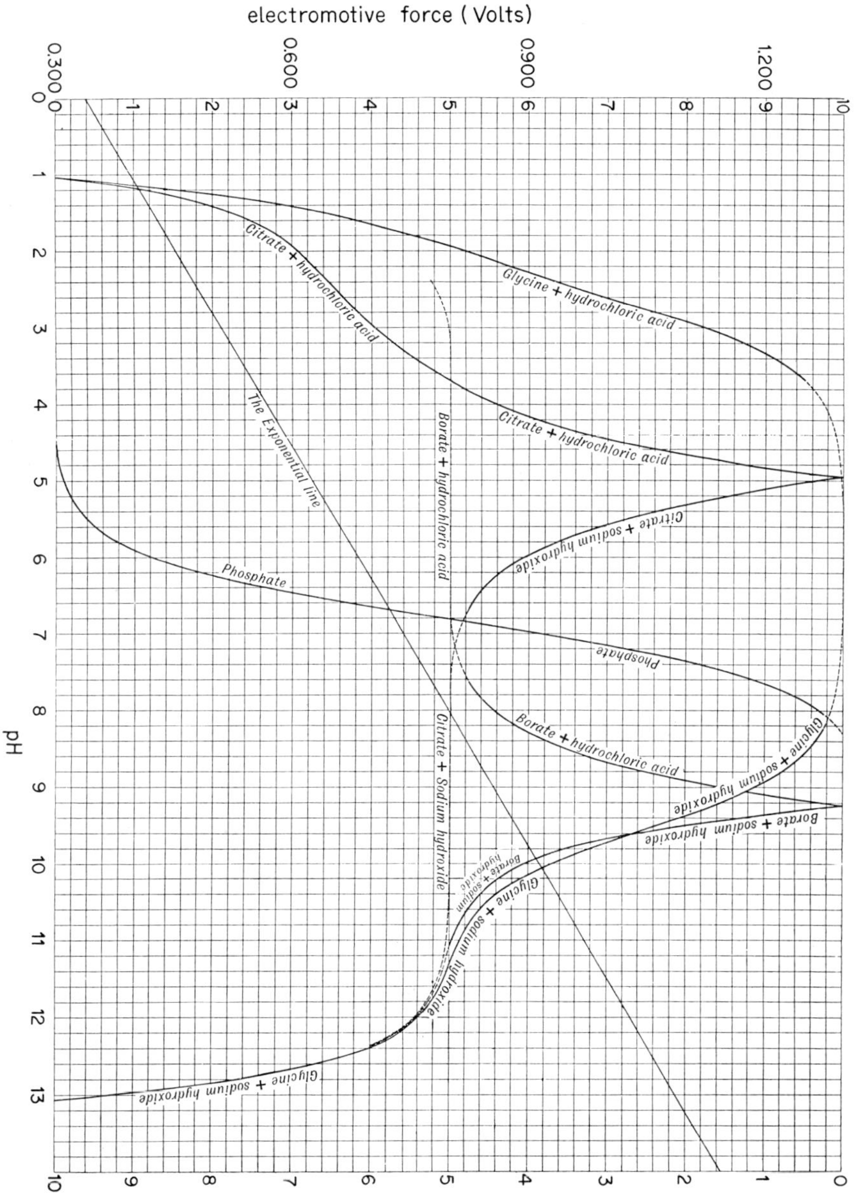

14 Chart of the Sørensen buffer system, which for many years acted as the standard for the definition of pH. The system consists of a succession of paired solutions, mixed in the quantities indicated by the ordinates to the right and left. The pH-values of the resulting mixtures are given on the abscissa.

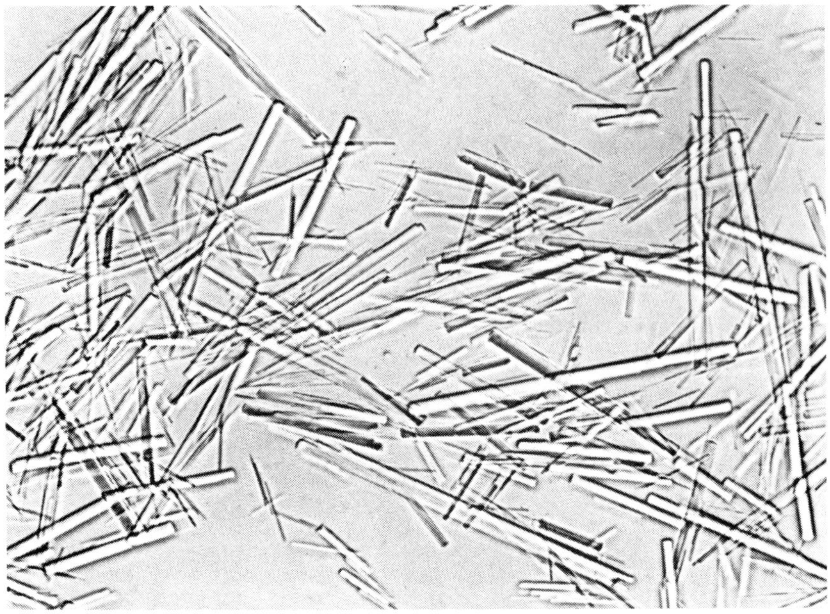

15 Egg Albumin crystals. The longest crystals are about 0.2 mm in length.

the dissociation of electrolytes, which made his calculations less significant. Nevertheless the comparison of the acid and base binding curves of glycine and glycylglycine with those of egg albumin is very important and qualitatively very informative. A later work (Sørensen, Linderstrøm-Lang, and E. Lund 1926) attempted a more up-to-date treatment of the subject, roughly describing the facts about egg albumin.

The conditions governing the solubility of egg albumin were also made the subject of a thorough study in Vol. 12. Sørensen realized that in such investigations it is very important to define as accurately as possible the two phases, the mother liquor and the crystals. He began therefore with an investigation of the composition of the crystals and for this purpose used his »method of proportionality«. It appeared that the crystals contained both water and ammonium sulphate but that the quantity of water was much greater in proportion to the quantity of

16 Sørensen crystallizing the protein Egg Albumin by ammonium sulphate precipitation (1936).

ammonium sulphate than was the case in the mother liquor from which crystals had formed. By gradually crystallizing egg albumin from an ammonium sulphate solution, the composition of the mother liquor was therefore changed as proportionately more water than salt was removed with the crystals. In this convincing way Sørensen explained the results obtained by H. Chick and C. J. Martin according to which the solubility of egg albumin in ammonium sulphate decreased with increasing quantity of crystals and he showed that the solubility of egg albumin was essentially independent of the quantity of crystals when the composition of the mother liquor was constant. This demonstration of the validity of J. W. Gibbs' Phase Rule for such colloidal systems was again pioneer work, the significance of which has not been lessened by time.

The last part of Vol. 12 treats of the osmotic pressure of egg albumin solutions. Here too Sørensen rejected the view put forward by colloid chemists that it might lead to erroneous conclusions if the properties of solutions of the colloid substances were compared with those of solutions of »crystalloids«. He felt convinced that the same »laws of nature« must apply in both cases and that the osmotic pressure of an egg albumin solution ought to be measurable and definable no less than the osmotic pressure of a sugar solution (even though with somewhat greater difficulty). It is now easier to be just to both these views and similarly to understand why Sørensen backed the right horse. Time has proved him right to a degree far beyond that which he himself imagined. The word »molecule« has entirely superseded the word »micelle« in the chemistry of proteins. On the other hand, time has dealt rather less kindly with the details, especially the theoretical details, of this very exceptional experimental work; but that does not detract from its monumental nature and originality. It has played an exceedingly important part in subsequent investigations.

Then came the end of the Great War and the difficulties of obtaining material. Two years passed with the work at a comparative standstill during which a frightful amount of time was taken up in preparing substances which it was not possible to buy. At this time Sørensen's interest in the practical application of his scientific achievements began to awaken and he became an adviser to many industrial concerns. Gradually during the course of the next 20 years more and more of his time was taken up with such activities, and he handed over much of his scientific work to his pupils. However, he did not lose his hold as the leader of the work of the laboratory. The change was mainly noticeable in that, to a greater extent than hitherto, he turned over to his students the responsibility of publication of the completed works which it became difficult for him to find the time to write up. But important papers continued to issue from his pen during this period. These works with those of several of his pupils were collected by him in 1930 in the treatise »The Constitution of Soluble Proteins« (»Reversible Dissociable Component Systems«). In this work Sørensen again was guided by his firm belief in the applicability of simple chemical laws to colloidal systems. His starting point was his investigation of the solubility of serum globulin (Protein Studies 8 (1923)) which, in agreement with earlier experiments by W. B. Hardy, proved that the solubility of this substance in dilute salt solutions increased with increasing quantity of the amorphous solid or liquid phase (designated »precipitate«). An indication of this fact had already been met with in egg albumin and was in that case explained by the presence of a small quantity of soluble, non-crystallizable, but coagulable protein in the crystals. Sørensen interpreted the globulin experiments in a similar way assuming that the globulin consisted of several components having different solubilities which explained the apparent deviation from Gibbs' Phase Rule. Furthermore, he went a step further and assumed that the different components, in the solution as well as in the precipitate, entered into loose chemical combination with each other, forming complexes. On altering the salt concentration ect., those complexes could dissociate into other smaller complexes and again re-form when the experimental conditions were re-adjusted to those prevailing at the outset of the experiment. Time seems to have proved that Sørensen was correct with regard to both these assumptions. Certain proteins were successfully separated into fractions having different chemical compositions (serum globulin, serum albumin, casein, gliadin, (G. Haugaard and A. H. Johnson (1930)), and later experiments by T. Svedberg and his collaborators indicate that the molecules of proteins, when subject to various influences, can be reversibly dissociated into smaller molecules. However, it was easy to criticise, from a theoretical point of view, the path by which Sørensen arrived at this the second of his postulates and on the basis of his own experiments he did not succeed in giving any proof of its correctness. This fact nevertheless does not prevent us from admiring the intuitive vision with which he maintained his view-point. With regard to his first assumption he stood on firmer ground and in that respect his work has had immeasurable pedagogic significance.

During his last years when he worked most of the time with his wife, Margrethe Sørensen (née Høyrup), he did much important work such as investigating the solubility of carbon monoxide hemoglobin in ammonium sulphate and carrying out a large scale investigation of the protein in whey.

Briefly stated it is this work with which Sørensen's name is direkctly connected. To all this must be added the numerous activities and work which he has personally influenced in his capacity as adviser but which it would take too long to relate here. We who are proud of having been Sørensen's pupils remember with gratitude what he taught us and see in our own work, where it is best, deep traces of his scientific personality. In this the many scientists from all over the world who have visited the Carlsberg Laboratory will wish to associate themselves with us.

The unrecorded part of Sørensen's work, his practical work

17 Sørensen titrating, using the original set of buffer stock solutions and indicators (1936).

of service outside the laboratory, deserves a whole chapter to itself. He had the joy of seeing his scientific work put to immediate pratical use. In the chemical industry, in all the industries in the daily running of which biological processes are employed, fermenting, baking, tanning, etc., advance followed in the wake of scientific progress.

My old, now long departed, friend Hans Jessen-Hansen who was a coworker of Sørensen's for 30 years, used to say that Sørensen's work had given the Carlsberg Brewery excess profits that many times covered the expenses of running the Laboratory. That is probably true, and I believe the managers of many other industries in Denmark will agree with me when I say that Sørensen's scientific work had the same positive value for them; but it goes without saying that his direct intervention as adviser – and naturally this was often necessary – had just as great, perhaps even greater signifiance. I myself had the joy of late years of following one branch of his advisory activities in industry, namely at the Carlsberg Brewery, and I have marvelled at his unimpaired insight and his delicate power of negotiation. At times he complained at the burdens that were thus laid upon him but his complaints were never very convincing; actually he loved to take a hand in the practical application of science, for he could do it so well.

His position in public life*), I do not feel called upon to speak about. He often talked with me about that side of his life but not always with exactly gentle words concerning the wire pulling of professional politicians. However, he was heartily glad to meet men to whose opinion he could bow in understanding; moreover there may have lurked in his heart a slight fondness for that interplay among men. He often remarked with a tolerant, mischievous smile: »It will come all right but it must go through the mill«.

As a research worker Sørensen was an experimenter. His investigations all lay within the sphere of preparative, analytical, and synthetic chemistry. Supported by his exceptional

* A consequence of his position in industry in general, and in later years especially his activities in the Danish Society for Civil Defense against chemical and air warfare.

memory, methodical habits, and organising ability, he was able to master a great amount of experimental material and thereby exhaust his subject at one go. His critical ability was outstanding. He was a past master at correcting errors in method and if any came to him with strange results, they left him with capital advice on how to improve the plan of the experiment. As already mentioned he seldom theorized, – he himself said that he lacked the ability so do so, – and he therefore always first asked if the experiments were correct before he began to philosophize on the significance of the results. We see this attitude of his throughout all his best works. Sustained by an unshakeable faith in the theories of physical chemistry for which he cherished a touching esteem, but to which he seldom dared add any independent contribution, he fought, with a highly developed gift for experimenting, to show that all strange experimental results which would not harmonize with these theories were due to the faulty arrangement or execution of the experiments.

However, his theoretical insight was considerable, – his gifts were by no means confined to his hands. He possessed the mental quality which, when undeveloped is called common sense, but when mature as in the case of Sørensen, is termed intuitive vision. He knew how to deepen and to confine, to reject and to simplify, and his work bears therefore the stamp of a rare completeness.

Sørensen loved work for its own sake. I have never seen him dispirited. About 1920 when I came to the laboratory he planned every day in detail what each assistant was to do, giving us in the morning his instructions written on a sheet of yellow foolscap in a neat beautiful hand. Later on when other activities took up more of his time and in addition he felt we had been drilled sufficiently, he talked with us instead. His ability to arrive concisely and exactly at decisions on the varied problems which he encountered during his rounds of the laboratory, always seemed to me to be exemplary. In later years when he spent so much time on the telephone, this ability of his to change the subject abruptly was invaluable. One minute in his office he could be discussing the cultivation of potatoes on the telephone, and the next, after replacing the receiver, turn with a laconic: »Well, and what is on your mind?« and enter brightly and fully into the discussion of any scientific problem presented to him.

That anyone from mere passing fatigue could develop even a momentary aversion toward chemistry in general was to him utterly incomprehensible. Science and its employment in the service of mankind was his only passion, it was the principle on which he built his life. I never came across any scientific question that did not interest him and which he was not willing to hear about however busy he might be. Therefore he was the most encouraging chief imaginable. He rejoiced with his fellow workers when their experiments went well and when a work was finished he would read it thoroughly, once quickly in order to get a general impression and once slowly and critically; thus not one single printer's error escaped him. He said he owed it to the laboratory.

He was a wonderfully industrious man and liked one to work overtime as he himself did. On Sundays he always went the rounds of the laboratory and those he met were greeted with a characteristic: »So you are in the workshop, are you?« Then he would take time to chat and many of my dearest memories of him come from these quiet Sundays when we wandered up and down by the hour talking of everything under the sun that interested him.

As a human being he was very sensitive but endeavoured to hide it. As a child he had had a highly nervous disposition which threatened to come between him and the world he wished to conquer. Attacks resembling epilepsy followed any strong psychic impressions, and he stammered. Both afflictions soon disappeared however. The stammering he himself cured by taking solitary walks in the woods around Sorø where he was at boarding school, and the attacks ceased by the time, when at the age of 18, he came to Copenhagen. For him work was the great healer and it is not surprising that he let it fill his life completely. Art, literature, and music he might be fond of, but mildly and at a distance. He had to protect himself against everything that was likely to be upsetting. Harrowing fictions such as are found in books, disturbed the mental balance he had won and were rigidly banned. Where real life was concerned however he knew a more beautiful way of disengaging himself from mental distress. No one so willingly and often helped unhappily placed people as he; a constant stream of unemployed passed through his office; sealed envelopes containing a ten kroner note were put aside for times when he was not there. He could not stand that anyone should suffer. In later years he was easily moved to tears when speaking of his protégées but he had an exceptional power of dismissing the painful and returning to his work, his deliverer.

There was also something very simple and trusting about him, an indomitable faith in humanity which, sad to say, was often abused but which time and again found its most beautiful expression in his attitude towards his co-workers and others who applied to him. I only knew him for 20 years, quite a span, but only the latter part of his life. Twenty years ago he could be severe over mistakes committed through stupidity or carelessness and he expected a lot from his pupils; but that is no more than every good teacher ought to do. Mild in his form of expression he ever was; both in his speech and in his scientific style he followed the motto: »Suaviter in modo fortiter in re«, which he often quoted. His even temper, his sincere smile, his fatherly interest in the festivities at the laboratory, also come to mind. Time made him gentler, he became a delightful old man, calm and wise. The many honours showered upon him pleased him but did not disturb his mental balance. Just, unselfish, and warm-hearted, he has performed his important work in the service of science and for the welfare of Denmark.

H. Schjerning

1862–1914

by *R. Koefoed**

H. Schjerning was born July 22, 1862 on a farm in Falster, the southernmost of the Danish islands. He lost his father when he was only 7 years old and together with a slightly older sister was brought up by his mother in Viborg in Jutland under straitened circumstances. There was not enough money to let the boy study medicine which was his intense wish. Instead he became an apprentice in a pharmacy, managed to take a pharmaceutical degree and started work as a chemist's assistant in Horsens. But his thirst for knowledge and further development induced him to abandon this position after one year. He went to Copenhagen, supported himself as a tutor in chemistry, and was soon in great demand. At the same time he studied chemistry and its auxiliary sciences. In 1886 he took a partial degree in chemistry at the Technical University. From 1887 to 1890 Schjerning held a position as assistant at the chemical laboratory of the Veterinary and Agricultural University in Copenhagen.

These years were, as far as work was concerned, probably the happiest in Schjernings life. He derived great satisfaction and joy from his teaching activities, and he found in Odin T. Christensen, the head of the laboratory, a good and understanding friend, much of his own kind in thought and temper. Schjerning was well aware of this situation but he felt nevertheless that he ought to accept when in 1890 the directorship of the laboratory of the Ny (New) Carlsberg Brewery was offered him. This position not only implied better economic conditions

**Written in 1915.*

18 The only available picture of H. Schjerning, taken in his later years.

but also the opportunity to engage – besides the daily control work for the brewery – in independent scientific investigations of the many puzzles which the fermentation industry still offers, in spite of much progress.

His new job began November 1, 1890, and continued until the merger between Gamle (Old) Carlsberg Brewery and Ny Carlsberg Brewery prompted Carl Jacobsen to abolish the position at the end of April 1904. On May 1st Schjerning became attached to the Carlsberg Laboratory. His position was that of Direktor Extraordinary under the Chemical Department. But since no working space was available in the new building of the Carlsberg Laboratory, the brewery placed part of the premises in the old Carlsberg Laboratory at his disposal, the very rooms in which Emil Chr. Hansen had carried out his early and fundamental work. Here Schjerning entered upon a succession of peaceful years of scientific work under relatively good working conditions. He was able to continue and, fortunately, bring to its conclusion the great project that he had planned and embarked upon in the nineties.

Most of his life Schjerning had to contend with weak health. He used a great deal of energy in not letting himself be subdued and only seldom was it noticeable to others than his closest associates that his strength failed and that his working capacity and productivity were impaired. For this reason it was also a surprise for most people when it was announced that he had died after only a few days of illness on November 9, 1914.

In 1894 Schjerning had married Emilie Broe who survived him. The marriage was harmonious and happy. There were no children.

Schjerning was a pupil of the inorganic chemist S. M. Jørgensen and was strongly influenced by Odin T. Christensen, Jørgensen's collaborator through many years. Not surprisingly therefore, Schjerning's experimental work during the first years dealt with problems of inorganic chemistry. After a small paper on analytical technique, in 1890 he published a paper on the chemistry of some manganese compounds, and i 1892 a paper on compounds between phenyl hydrazine and salts of the magnesium group.

But from then on Schjerning was exclusively occupied by problems of an entirely different kind. During his work in the brewery control laboratory he was often exposed to a special technical difficulty termed »glutin turbidity« which causes beer, that in all other respects is unexceptionable, suddenly to become turbid upon cooling either in the brewery or in the customer's storage. This turbidity is due to precipitation of a protein that cannot be redissolved. This occurrence could not be explained, so no means to remedy the deficiency were at hand although it often caused considerable trouble.

Schjerning performed a series of preliminary experiments from which he concluded that there was only one way to elucidate the phenomenon: one had to differentiate, if possible to characterize the various proteins as groups of related proteins present in barley; moreover, one ought to follow the transformation and breakdown of these proteins during the various processes that barley is subjected to in the brewing. It was a long and laborious path that Schjerning proposed to follow, but right from the beginning he designed a painstaking, well considered plan for the whole work. He followed this plan and to his great credit he managed through 20 years of perseverance to reach the end of the road.

Schjerning published the results of these investigations in a number of papers during the years 1894–1912 in various journals including the Carlsberg Laboratory's own journal, the Comptes Rendus des Travaux du Laboratoire Carlsberg, and after he had brought the series to its conclusion, upon special request from the Laboratory's Board of Trustees, he gave a comprehensive review of his work in the Laboratory's journal.

The first part of Schjerning's work was the development of a technique for the quantitative determination of the various groups of proteins involved. This technique was based on the empirical fact that certain soluble proteins can be quantitatively precipitated by defined precipitating agents.

With this method as his tool, Schjerning tackled his main

task and followed qualitatively and quantitatively the fate of the proteins of barley (as defined by his method) throughout the various processes of beer production. But he never forgot that he was dealing with biological processes and therefore had to expect considerable variations: therefore, he extended his investigations to barley samples from different years in order to obtain results of more general validity.

This work which is probably rather unique in its concept, has thrown light upon many matters previously unknown, and has given the beer brewing industry a safer basis for work. Among other things it elucidated the characteristics of barley as a »malting staple« and it has given guidelines for choosing the proper processes which ensure satisfactory results under the varying conditions of brewing practice. It should also be mentioned that Schjerning succeeded in answering the question which constituted his original point of departure, The glutin turbidity of beer was found to be caused by a protein »Albumin II« (later named Edestin) belonging to the globulins and displaying the characteristic property of this group: in dilute salt solutions (e.g. in beer) they are sooner or later transformed into an insoluble substance. By means of a large amount of experimental material, Schjerning reached the conclusion that the key to the solution of the turbidity problems was to be found in the employment of malt that does not contain »Albumin II«, in other words that the malting process has to be performed under conditions that ensure a suitable transformation of proteins.

After the completion of the work according to his preconceived plan Schjerning wanted to utilize the experience obtained in more profound investigations concerning e.g. the significance of the genetic characteristics of barley for the completion of the malting process. He also planned to isolate the various individual proteins of barley, but death interrupted his work.

Schjerning produced an overwhelming amount of experimental material which is correct beyond doubt and which will preserve its value in time to come. The conclusions he drew from that material must be interpreted in terms of the state of knowledge of proteins available to him at the time. But subsequent fundamental developments in the chemistry of proteins will undoubtedly result in changes regarding the interpretation of the experimental results.

As a scientist, Schjerning was characterized by faithful perseverance in pursuing an established goal through years of work; by the modesty with which he reported his results as they became clear, step by step; and by the honesty with which he meticulously acknowledged the contributions of others to a field that had become his own.

The same traits characterize Schjerning as a human being. He was a quiet man, and since he never pushed himself forward, he remained comparatively unknown, even among his colleagues. But his quietness, undoubtedly a result of the difficulties that he had met in his youth and adolescence, was far removed from indifference and half-heartedness. When he met injustice against others, against himself or just against what he deemed to be right, he gave battle without hesitation. And if he thought that he was able to support a cause, he took up work for the sake of the cause alone. He liked young people and was ever ready to assist them by word and deed – he felt an urge to fit them for the struggle that life, according to his own experiences, consisted of. But even if life had been a struggle for him, it had also brought him much happiness; and out of his religious background and thanks to an open mind for the beauties of nature and of human life, he preserved throughout his lifetime the capacity of feeling joy.

K. U. Linderstrøm-Lang

1896–1959

by *H. Holter**

The outward course of Linderstrøm-Lang's life was quite devoid of complication. His name will always be associated with the Carlsberg Laboratory, to which he came just after completion of his examinations and at which he remained until his death. Within this simple framework a great man and his work came to full expression.

Kaj Ulrik Linderstrøm-Lang was born on November 29, 1896 in Frederiksberg, a suburb of Copenhagen. His father, Carl Frederik Linderstrøm-Lang, was a teacher of German and Latin at the Frederiksberg Gymnasium and the son of the cantor at Vemmetofte, a home for unmarried ladies of rank.

Teaching was a frequent vocation in the Lang family; many of Lang's forefathers had been engaged in this occupation, and his aunt, Theodora, was the founder of Th. Lang's »Skole og Seminarium« in Silkeborg. Thus, an atmosphere of schools and teaching based on the tradition of several generations prevailed in the Lang home in Frederiksberg.

Lang's mother, whose maiden name was Ellen Hedvig Bach, was the daughter of P. J. Bach, a banker in Svendborg. There was an artistic bent in the Bach family which, however, had been prevented from manifesting itself by economic difficulties. Lang's mother was musical and wanted to be a music teacher; her brother was an architect. Lang himself never doubted that his artistic temperament was inherited from his mother, and the preoccupation with art established an early and lasting bond between mother and son.

Lang was the youngest child and had two sisters, both of whom became shool teachers. They are now living in Silkeborg

**Written in 1959*

19 K. Linderstrøm-Lang. In the words of the photographer, his assistant C. F. Jacobsen: »He looks at you as he used to when beginning to think about something you just said«. (1939).

where one of them, Karen Linderstrøm-Lang, is rector of Th. Lang's School founded by her aunt. I am indebted to this sister for much information concerning Lang's childhood home and his adolescent years.

The Lang home on Kong Georgsvej undoubtedly excelled in many ways the average Copenhagen middleclass home at the turn of the century, and many of its customs became an integral part of Lang's later life. When he remembered that his parents had themselves woven the cloth with which their furniture was covered, it was only natural that he should take a great interest in the artistic furnishings of his own home. As his father before him, Lang was fond of establishing domestic rituals which were carried out with care both in daily life and on festive occasions. Reading aloud, an activity which Lang pursued with much gusto and talent, was also a tradition in his childhood home.

The relatively few friends and acquaintances of Lang's parents belonged chiefly to academic circles. Many parties requiring much preparation were held, with table arrangements, humorous afterdinner speeches and verses playing important roles on these occasions. Such customs, too, were perpetuated by Lang, a fact which is known to all who have enjoyed his fabulous hospitality. Many cultural problems were discussed in the family, some with considerable ardour, whereas political subjects do not seem to have aroused much interest; on the other hand, the family's faith in Christianity was a source of conflict to Lang during his adolescence, but it was a conflict lacking deep roots. There were other problems, however, which apparently were of great importance to him. As a child he felt himself to be hopelessly ugly. His red hair, freckles, thin neck and bad posture gave him an inferiority complex and caused him to be self-conscious and unsure of himself as a youth.

The variety of his interests dominated this period to a greater extent than is usual among children, even highly gifted

20 Linderstrøm-Lang, 15 years of age, with his sister Karen in the garden of the family home.

ones. He drew, painted, modelled, wrote poetry and played the violin, though at the same time he was interested in carpentry and especially fond of making fireworks in a neighbour's hobby cellar. When he felt the need of an audience for his productions, he could always come to his mother or youngest sister. However, a real appreciation of his many-sided talents was lacking in the home. It was almost as though the parents, especially the father, were slightly in awe of the boy's ability and perhaps afraid that their praise might tend to spoil their youngest child.

Lang's school years at the Frederiksberg Gymnasium were normal in all respects, and none of the teachers seems to have had any special influence upon him, nor were they responsible for awakening his interest in science. His dissatisfaction with his personal appearance caused him a good deal of grief; also, it was not altogether easy for him to be the son of one of the school's teachers, author of a not too popular Latin grammar.

The decisive event of this period was his father's swift death when Lang was 15 years of age. Suddenly, everything was changed. His mother was left in a difficult economic situation; the house on Kong Georgsvej had to be sold, thus altering their whole way of life. But the most tragic shadow was his mother's long and painful illness. She died nine years after her husband of tongue cancer, for which she had one operation while her husband was still alive.

Lang completed his secondary schooling in 1914, and at the age of 17 he began the study of chemistry at the Technical University in Copenhagen. These student years played an important role in his life, though more with respect to personal contacts than to knowledge acquired. He was highly esteemed by his fellow students as playwright for their annual theatricals, as editor of »S. K.«, the chemical students' journal, and for many similar achievements. Actually, Lang was not overly enthusiastic about chemistry which he regarded primarily as a means of earning a living. He managed to get through his studies, bot no more, for his true interests lay in other fields. He was possessed by a powerful urge to express himself in artistic creation, but the multiplicity of his talents prevented him from deciding upon a definite course. As writing interested him most, it became a necessity that he determine the scope of his ability as an author. He was bold enough to send one of his plays to Georg Brandes, who replied: »Your manuscript has the virtue of not being boring. Whether or not you have talent is something you should best be able to decide yourself.« Lang was none the wiser for this oracular reply, and probably more decisive, from a practical point of view, was the fact that no publisher would accept any of his work.

Lang could find no solution to his problems, and in the last years of his studies his restlessness culminated in a serious crisis, further accentuated by an unhappy love affair. His mother, worried that he might abandon his studies altogether, arranged for him to leave Copenhagen for half a year. At the end of this time, which he spent on a farm in Jutland, he returned home with a suitcase full of plays, in a fine frame of mind, and willing to take his examinations.

In 1919 he received his degree after having passed his examinations, if not with flying colours, at least in a very respectable fashion – a fact which is noteworthy in view of what had passed.

In the spring of 1919 Lang obtained a temporary position at the National Research Institute of Animal Husbandry in Copenhagen in its department of chemistry. The head of this department, A. C. Andersen, who had previously been at the Carlsberg Laboratory, soon realized the extent of Lang's ability and mentioned his name to S. P. L. Sørensen. Professor H. Baggesgaard-Rasmussen, who had known Lang at the Technical University, also spoke well of him. Furnished with these recommendations, Lang applied for a position as assistant in the Chemical Department at Carlsberg and obtained it in August, 1919. He always claimed that he had wanted this particular position not for the sake of science, but because he thought it would give him much free time for writing. This was, of course, meant as a joke, but I believe there was a good deal of truth in it. Though he had at this time practically surmounted his crisis, it only became completely a thing of the past when science took hold of him, and it is very doubtful whether this would have happened had he come to any other man than S. P. L. Sørensen.

At this time Sørensen stood at the height of his career. Practical work in the laboratory was, however, very much curtailed by the extreme shortage of necessary chemicals in this period just after the war. Lang was put to preparatory work, and Sørensen's experimental talent together with his insistence on purity resulted in an intensive period of training in practical laboratory chemistry. This type of work interested him, even though each morning when he arrived there lay on his desk a note, written in Sørensen's meticulous hand, containing detailed instructions for the day's work.

Contact with a man to whom science meant absolutely everything, and whose extraordinarily successful life had been built solely upon that foundation, was just the kind of influence Lang needed. From Sørensen he learned not only biochemistry but how to become a scientist.

In addition to the personality of Sørensen himself, the situation at the Carlsberg Laboratory was favourable for a young man with talent. A number of Sørensen's more prominent staff members had left Carlsberg shortly before Lang's arrival, creating a vacuum which Lang successfully filled. He soon became Sørensen's trusted assistant.

Lang was married a few years after joining the Carlsberg Laboratory. Quite in accordance with the family tradition his wife, Gerda Kyndby, was a school teacher from the Th. Lang's institute in Silkeborg, and was also his sister's best friend. She was the daughter of Johannes Kyndby, a school teacher in Herning. The wedding took place in March, 1922, when Lang was 25 years of age. Three children resulted from this marriage, two daughters and a son, and as the housekeeping and representative duties gradually expanded in scope, Mrs. Lang became a very busy person indeed. But she never gave up the teaching profession which she truly loved and which her husband appreciated and respected.

Lang's first scientific contributions were of a physical-chemical nature. In 1920, Einar Biilmann had described the quinhydrone electrode, and Sørensen, who was naturally interested in employing this new electrode in his hydrogen ion measurements, requested Lang and his own wife and assistant, Margrethe Sørensen, to investigate how it would work out in practice. It was soon shown that, in the presence of salts, quinhydrone measurements deviated from measurements carried out with a hydrogen electrode. A description of this salt error was the subject of the first scientific paper bearing Lang's name; it was published in 1921, together with S. P. L. and Margrethe Sørensen.

The problems entailed by this work were followed up in a longer theoretical investigation which Lang himself considered his debut; this was entitled »On the salting-out effect« and appeared in 1923. In its experimental section, this paper dealt with the influence of salts on the solubility of hydroquinone, quinone, succinic acid, and boric acid. In the theoretical treatment, Lang went considerably beyond the immediate problem in a general discussion based on the activity theory of G. N. Lewis, N. Bjerrum and J. N. Brønsted, and especially on the hypotheses of P. Debye which at that time were quite new.

I think it can be said that this paper was for Lang much more than a debut. It marked his conversion to science, and from that time on there was no doubt in his mind. The work had involved a detailed study of the classics of physical chemistry, of which J. W. Gibbs' contribution was his favourite, and the validity of Lang's viewpoints was tested in many discussions with Niels Bjerrum, J. A. Christiansen, and Holger Jørgensen. Lang often mentioned to me that these discussions were of great significance in his development, and it is therefore understandable that, although he was working in a biochemical laboratory, he actually considered himself to be a physical chemist. He felt that his vocation was to supplement S. P. L. Sørensen's practical and experimental work with a mathematical and speculative background, which he thought was somewhat lacking in the laboratory.

In 1923 P. Debye and E. Hückel's famous theory of electrolytes appeared, and only one year later Lang, on the basis of that theory, published a classical work entitled »On the ionisation of proteins.« In this paper he extended Debye and Hückel's conceptions and adapted them to the conditions prevailing in solutions of polyvalent ampholytes, as known from Sørensen's investigations, and showed how the shape of the titration curve varies with the ionic strength. To this day Lang's work forms the basis of all studies of protein titration curves.

The first application of these ideas is found in a contribution which was published in 1926 by Lang and Ellen Lund on »The influence of salt concentration on the acid-binding capacity of egg-albumin.« This paper contains, among other things, the first formulation of the important difference in definition between the iso-electric and iso-ionic reaction of proteins.

The problems which were opened up by these studies occupied Lang for many years. He followed the development of the field very carefully, even when he was engaged in other problems, and whenever he felt the urge to wrestle with mathematical theory, he would take up some problem dealing with protein charge. Several of these attempts ended unpublished in his desk drawer; a finished work in 1953 on »The activity coefficient of large multipolar ions« deals with the very difficult mathematical problems concerning rodshaped and ellipsoidal ions in media of varying ionic strength. Just before his death he completed, together with S. O. Nielsen, a review of the whole field under the title »Acid-base equilibria in proteins.« This paper was published in 1959, and shows clearly that he was still very much interested in the first love of his youth.

But to return to Lang's first years at the Carlsberg Laboratory, still another important contribution deserves mention. This was a paper in 1926 entitled »On the state and stability of colloid solutions.« This study stands somewhat isolated among Lang's productions, but it is actually very characteristic of its author. As previously mentioned, he had been intensely interested in the thermodynamic studies of J. W. Gibbs. Since he was never content to be merely receptive, always desiring to apply acquired knowledge to a problem of his own, he produced a detailed thermodynamic treatment of the abnormal osmotic behaviour of colloidal solutions. On such occasions Lang was inclined to let the charm of mathematics coerce him into expending more effort on a problem than the results could justify. This study, which he himself in later years did not value very highly, at least gave him the opportunity of long consultations with Niels Bjerrum. These discussions were the nearest Lang came to being a disciple of Bjerrum, a scientist whom he held in the highest esteem throughout his life.

In addition to the above mentioned work Lang also participated in investigations which were of special interest to Sørensen – namely, the factionation of proteins. While Sørensen and his wife investigated the serum globulins, Lang and a Japanese guest, S. Kodama, took up the question of the heterogeneity of casein. This resulted in three publications of which the last, published in 1928, was Lang's doctoral thesis.

Today it is difficult to understand the uncertainty which prevailed thirty years ago concerning the nature of the more complex proteins. Lang's contribution was important, showing, in contrast to the generally held views, that casein was not homogeneous but a mixture of several closely related components. These were separated by a fractionation procedure which at that time was very difficult; when the components were mixed in the right proportions, they yielded again the original product. This work, carried out with all the precision which Lang had learned from Sørensen, was a contribution to a discussion which today is chiefly of historical interest. Modern methods of protein chemistry have progressed so rapidly that Sørensen's views, in support of which Lang's work was undertaken, today appear only as a stepping stone.

Already at that time Lang was aware of this development. Despite his veneration for Sørensen, he was not in complete agreement with his concept of proteins as »reversible dissociable component systems«, and this certainly contributed to the fact that Lang took up other problems as soon as he had obtained his doctor's degree.

The field which attracted him was that of proteolytic enzymes. But first I must mention a by-product of his doctoral dissertation – the determination of amino groups by titration in acetone with naphthyl red as an indicator. Lang needed this method in the analysis of nitrogen distribution in casein fractions. His familiarity with Bjerrum's and Brønsted's acid-base theories aided him in the selection of a suitable non-aqueous medium – 90% acetone – and a weak basic indicator, which together constitute the keypoints of the method. The method was tested in the customary Sørensen manner on the laboratory's entire collection of amino acids, and in the customary Lang manner the introduction to the paper contained an admirable discussion of the whole theory of titration. He now had an excellent tool for measuring proteolytic action, and the technique made his name known among enzyme chemists. Many foreigners, including myself, first came across the name of Linderstrøm-Lang through the use of the Linderstrøm-Lang acetone titration.

Lang's interest in proteolytic enzymes was aroused in 1926, and he received a fellowship in order to travel to München to study the methods of the R. Willstätter school. This trip was of only two months' duration, and the Geheimrats-atmosphere in Willstätter's laboratory was a source of considerable wonder to him, but he managed to learn what he had come for and was, of course, pleased with his first scientific journey.

However, Lang never became a supporter of the German

school of enzyme chemistry. Instead, immediately after his return to Copenhagen, he began to lay the foundation of his own.

In the late twenties, the Chemical Department of the Carlsberg Laboratory was visited by many Danish and foreign scientists, but Sørensen, whose fame naturally was the chief cause of these visits, was becoming at this time increasingly occupied with his interests outside the Laboratory. It was therefore quite natural that the guests in reality came into contact mostly with Lang, and after his return from München the whole laboratory was increasingly permeated by his ideas. Those visitors played a very important role for Lang, both scientifically and in private life. Contact between him and his colleagues was a true exchange, hearty and fruitful, and the complete antithesis of any Geheimrats-atmosphere. Many ties of friendship were established which developed into a network over the whole world and which contributed greatly to the fact that Lang was known and welcome whereever he happened to go.

The series of investigations on proteolytic enzymes, which was initiated in the late twenties, was soon centered on the problems surrounding the definition and specificity of the peptidases. At that time, dipeptidases were distinguished from polypeptidases, and it was believed that specificity was dependent upon the length of the substrate chain. Lang showed in 1929–30 that two dipeptides, leucylglycine and alanylglycine, were attacked by two different enzymes, and that one of them could also break down longer peptide chains provided that they contained a terminal leucine. Thus it was not the length of the chain but the configuration of the substrate which was significant. Soon after publication these studies were severely criticized, especially by the »German school«, and a considerable period elapsed before it was proved that Lang's views were correct and that a nonspecific »dipeptidase« does not exist. Twenty years later, Joseph Fruton, one of those who have contributed most to the elucidation of the specificity of the proteolytic enzymes, delivered a lecture in Copenhagen, on which occasion he characterized Lang's early work on the peptidases as follows: »He was far ahead of his time, and first now the rest of the field is beginning to catch up with him.«

This research on enzymes initiated a shift in Lang's interest to the more biological side of biochemistry. This interest was stimulated by discussions which Lang and I held, after I came to Copenhagen in 1930 as one of the many guests of the Carlsberg Laboratory. We talked about a vague dream of mine, to attempt to trace the biological role of enzymes by somehow relating their activity to the structure of the tissues and cells in which they occur, since structure in some way or another must be related to the function of the enzymes. Lang immediately pointed out that an enzymatic mapping of tissues and cells would only give useful information if one worked quantitatively, and that quantitative determination of enzymes on the histo-chemical scale would certainly be beset with difficulties. Micromethods as yet non-existent would be necessary, but they could probably be devised if one were not afraid of working and planning far ahead.

The collaboration initiated by these discussions lasted for twenty years, and the technical problems which were forced upon us interested Lang so much that during the first ten years he devoted to them most of his energy and his unbelievable experimental ingenuity.

The group of methods developed first comprised microtitrations for estimating protein-, carbohydrate-, and fat-splitting enzymes in quantities approximately 1000 times smaller than could be determined with the then current macromethods. This scale was in no way sufficient for enzyme determinations in single cells, but it permitted us to work with tissue samples which were so small that their histological structure could be controlled under the microscope.

Despite the microscale, the main difficulties inherent in such

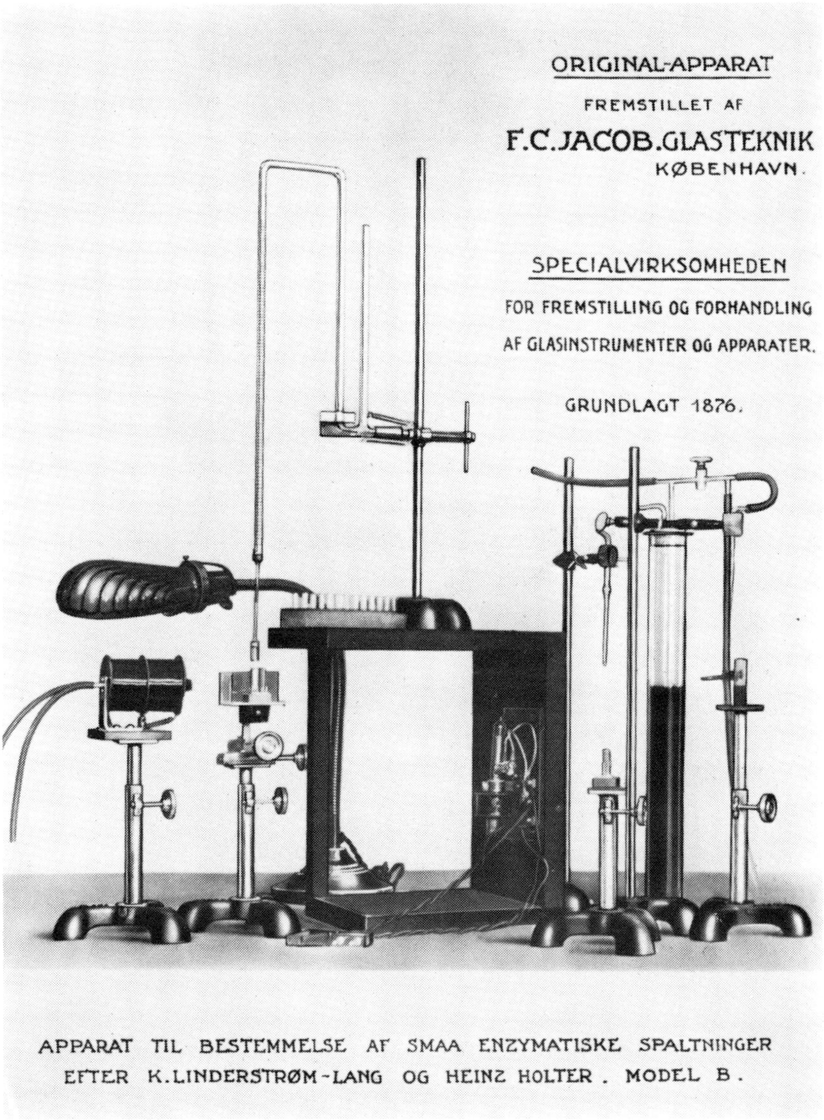

21 The original Carlsberg microtitration outfit. Page from the catalog of the Danish precision glassblower F. C. Jacob.

work are not encountered in the analytical procedures. The real difficulties are in the sampling, and especially in finding a reliable physiologically significant procedure for the correlation of histological and chemical data. It was most definitely

Lang's acumen in the solution of these problems which was responsible for the success of these investigations in creating a new approach to histochemistry.

These studies were received with much interest. The micromethods were so obviously useful, not only in histochemistry but also in many other branches of biology and medicine, that the laboratory was more than ever frequented by visitors desiring to study the new techniques. Lang was invited to give lectures on the subject at various congresses, and he enjoyed these opportunities of making contacts with biologists and members of the medical profession.

The genuineness of Lang's interest in biology was shown by the fact that when awarded a Rockefeller fellowship he chose for his studies the laboratory of the geneticist, Thomas Hunt Morgan, at the California Institute of Technology in Pasadena. This group was naturally predominantly biologically-minded, and Lang chose it expressly for the purpose of receiving new incentives. Scientifically, the benefits of this trip were not overwhelming, especially since Morgan himself was ill most of the time, but it was most rewarding from a personal point of view. On the way to and from California Lang visited a large number of laboratories – so many, in fact, that the otherwise hardened officers of the Rockefeller Foundation shook their heads, but Lang's faith in his own ability to establish personal contacts was completely justified. Many friendships which he made on this trip lasted a lifetime, and it was on this occasion that he laid the foundation for the enormous popularity which he later enjoyed in America.

Contact with the Rockefeller Foundation was not limited to this trip alone. With the passage of time it became even closer and in 1937 Lang received a grant in support of his investigations which was still running at the time of his death.

Sørensen was delighted with Lang's success, which was confirmed in 1935 when Lang was elected to the Royal Danish

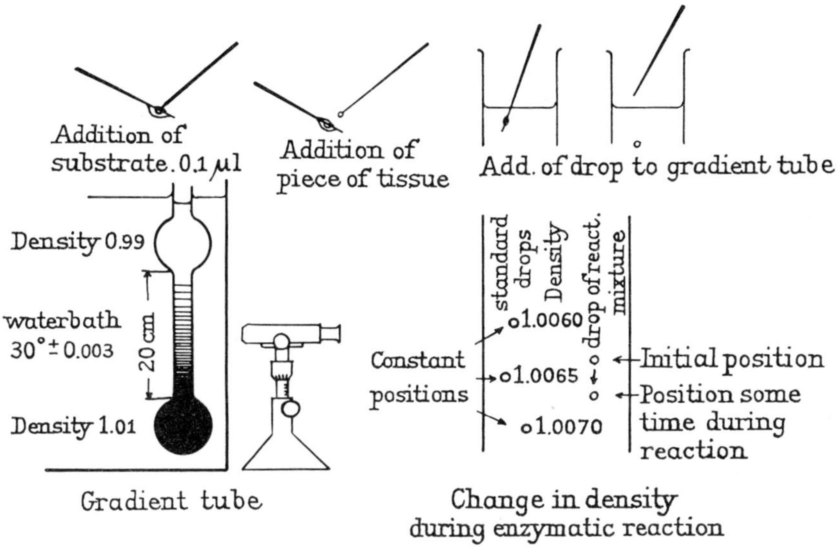

22 Micro-dilatometric technique. The »gradient tube« contains a mixture of kerosene and brombenzene of graded density. The substrate drop on the glass needle is surrounded by a drop of kerosene.

Academy of Sciences and Letters as its youngest member. As the time for Sørensen's retirement approached, it was a great reassurance for him to know that the man whom he himself had chosen as his successor had already obtained so much recognition that there could be no doubt of his appointment. Sørensen could thus contemplate the future with confidence.

On the first of April 1938 Lang replaced Sørensen as Professor and Head of the Chemical Department. A great man was succeeded by an even greater one, and there was no doubt that the laboratory would flower in the times ahead. With respect to the actual work of the department, the change was of little consequence, since it had long been influenced by Lang's methods of conducting scientific research. For Lang himself, the move into the official residence on Gamle Carlsbergvej meant much. At last he had found the environment which corresponded to his hospitable and sociable nature; the obligation to act as host for many guests was for him no burden but rather an opportunity of letting others share in his pleasures.

In 1937, the year before his appointment, Lang published two papers which decisively demonstrated his ability to develop fundamental techniques by simple means. One of these papers deals with the Cartesian diver, then only known as a philosophical toy, and describes its use for microgasometric measurements. The other concerns a method of determining the specific gravity of very small droplets of liquid floating in a medium of graded density, and the employment of this method in the measurement of those many enzymatic reactions that cause a change in the specific gravity of the substrate. Both methods were significantly (about 100 times) more sensitive than the previously mentioned titration methods, and consequently a whole series of new enzymes was rendered accessible to micro-analysis, on a scale which now permitted working with single cells or cell fragments rather than tissues. However, the most striking characteristic of these methods was their originality, Lang's ability to see how simple principles could be utilized as the foundation of whole new techniques which later proved to be so extraordinarily fruitful. I am sure that it is especially these two methods which J. A. Christiansen had in mind when he spoke about their »noble simplicity«.

Both methods, of course, have their complications and sources of error, and especially the Cartesian diver requires a good deal of insight into physical problems, such as diffusion, solubility of various gases in liquids, etc. With great enthusiasm Lang set about to investigate these problems and published a series of papers which culminated in 1943 with the monograph »On the theory of the Cartesian diver microrespirometer«. This makes difficult reading, especially for the unfortunate biologists who naturally should profit most by a careful perusal of it. If the truth were to be told, this treatise has not been read nearly as much as it deserved, simply because it is too difficult. Lang himself later admitted that »some of the sparrows could have been shot with smaller cannons.«

The elaboration of the diver theory was an outlet for Lang

during the troubled and scientifically restricted years of the second World War. The Laboratory had, of cource, no visitors from abroad; on the other hand, there were many other things to occupy our thoughts. Lang participated in the resistance movement with heart and soul; he established good connections with the illegal routes to Sweden, and many persons have him to thank for their safety. In 1944 he was arrested, together with his daughter's fiancé, Peter Fyhn. The Germans could not prove that Lang was guilty of anything and treated him with comparative leniency, but they nevertheless retained him in prison for fourteen days until intervention from various sources resulted in his release. But all efforts to save Peter Fyhn were in vain. He was shot in April 1945 as one of the last victims of the occupation. This was an extremely hard blow for Lang which in later years led to his reluctance to discuss his experiences during the occupation.

With the end of the war and the scientific isolation Lang's years of devotion to enzymatic histochemistry were also drawing to a close. In 1943 the Carlsberg Laboratory had been extended by a cytochemical sub-department, of which I became the head, and Lang decided to entrust the histochemical and cytochemical work to this new department. He himself would devote his time to that field which had always been closest to his heart – namely, protein chemistry. Thus, he himself never came to apply the Cartesian diver to biological problems in the same way as the titration methods. On the other hand, Lang continued to employ the gradient method, no longer for microdetermination of enzymes but for the study of the exchange of hydrogen with deuterium in proteins.

Lang's return to protein chemistry had already been heralded by three minor papers published during the years 1938–41. The first of these, »Peptide bonds in globular proteins,« was published together with R. D. Hotchkiss and Gordon Johansen. It concerned the hypothesis that a state of equilibrium exists between native and denatured protein in protein solutions. If enzymes attack the peptide bonds which are exposed by denaturation, the equilibrium is displaced and new protein can be denatured.

The two other papers, both published in collaboration with C. F. Jacobsen in 1940 and 1941, dealt with similar problems. One treats of methylthiazolin, a relatively simple sulphur-containing compound, as a model of the »masked« thiol-groups of proteins.

The second paper, on »The contraction accompanying enzymatic breakdown of proteins«, dealt with the volume contraction during protein splitting, and demonstrated with the aid of a very sensitive dilatometric method that the hydrolysis of the first peptide bonds in certain proteins is accompanied by much greater changes in volume than in the later phases of the breakdown, where the increments per peptide bond become normal. Lang concluded that the rupture of the first peptide bonds causes very great changes in the whole protein molecule, and he related these changes to denaturation and the well-known fact that many proteins are attacked by proteolytic enzymes only when they are in the denatured state.

This concept proved to be extraordinarily fruitful and led Lang into that field with which he was to be chiefly occupied during the rest of his life – the study of protein structure on a higher plane than the peptide chains, the study of the factors which stabilize these structures, and their changes during denaturation and proteolytic breakdown.

In this field Lang was completely in his element. It brought forth a synthesis of everything with which he had previously been occupied: his knowledge of proteins and enzymes, his mathematical ability, and his experimental ingenuity. This, combined with his wealth of ideas and clear reasoning, brought to maturation a work of exceptionally high quality. No wonder that his fame in the years following the war increased to heights reached only by a few.

His views on the mode of action of proteolytic enzymes were presented in two important review lectures, the one delivered at a Cold Spring Harbor Symposium in 1949, and the other in 1953 at a Solvay Congress in Brussels. The latter contains a detailed mathematical analysis of a series of theoretically possible reaction sequences.

An important sideline of the studies on protein structure and enzymes was disclosed by the observation that bacteria-infected ovalbumin preparations crystallized in a new crystal form – as plates rather than needles. Lang investigated the formation of the new protein, which was designated plakalbumin, together with Martin Ottesen. They found that the transformation of ovalbumin to plakalbumin is caused by a proteolytic enzyme which is produced by the infecting bacterium, and that the splitting which occurs is a quite superficial attack on the ovalbumin, resulting in the liberation of only a few peptides; the main part of the molecule is transformed into the new substance, plakalbumin. Lang and A. V. Güntelberg compared the molecular weight of ovalbumin and plakalbumin with the aid of an osmometer which was a modification of J. A. Christiansen's osmometer of 1917. Lang took this opportunity to provide a theoretical treatment of some problems concerning molecular association, which he considered had not been sufficiently elucidated in earlier investigations.

This carefully studied example of »limited proteolysis«, as the process was called, is analogous to the formation of a whole series of biologically active substances, in which a relatively small change in the protein molecule also results in a complete upheaval of its structure and properties.

The structural problems involved in these studies constituted the object of a series of investigations which occupied the last years of Lang's life. In 1951 he delivered the Lane Medical Lectures at Stanford University on the subject »Proteins and Enzymes.« In one of these lectures he formulated his ideas on the structure of proteins at three interdependent levels. The

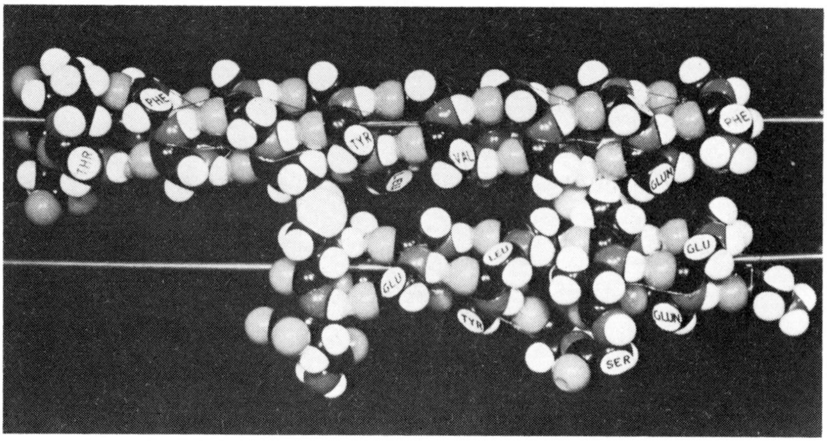

23 Linderstrøm-Lang's model of the hog insulin molecule, based on its »primary structure« and the deuterium exchange experiments. Years after his death the main features of this model were verified.

primary structure is the true peptide chain. The secondary structure stabilizes the peptide chain through hydrogen bonds in a regularly repeated pattern – a spiral or helix according to L. Pauling and R. B. Corey's theory. The tertiary structure comprises the bonds which occur between the various side chains of the amino acids – bonds which can both strengthen the spiral structure and fold it in a capricious manner, whereby the globular or oblong macromolecule appears which characterizes the respective protein in its undenatured condition. Any disturbance of this structural ensemble, whether at the primary, secondary, or tertiary level, will cause far-reaching changes in the whole molecule.

This terminology was rapidly accepted, and today it has become part of the everyday language of protein chemists; however, Lang felt that he must furnish additional experimental proof of his ideas by investigating the factors which stabilize the secondary and tertiary structures of proteins. For this purpose he chose to measure the speed with which hydrogen and deuterium are exchanged between a protein and the water surrounding its molecules in solution. The technique he em-

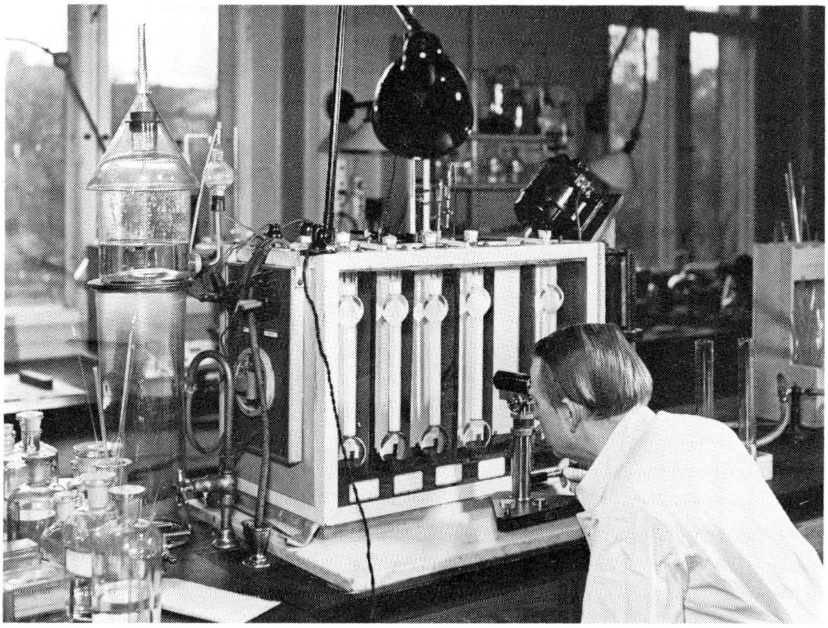

24 Linderstrøm-Lang performing a deuterium exchange experiment, reading the level of a micro drop in the gradient tube (1951).

ployed was an extension of the gradient method, measuring the concentration of deuterium by means of a specific gravity determination. This was described in 1954 and 1955.

In the years that followed, Lang investigated simple, non-folded peptide chains and showed that all hydrogen atoms which are bound to oxygen and nitrogen in such compounds are immediately exchanged. But in the more complicated proteins matters are different: only part of the O- and N-bound peptide hydrogen atoms can be exchanged, the others being made inaccessible by masking effects in the secondary and tertiary structures. In denaturation these structures are unfolded, permitting the masked hydrogen to be exchanged. By measuring the rate of exchange under varying experimental conditions, it is thus possible to determine the number of peptide hydrogen atoms involved in the secondary structure, and to measure their stability. This approach which in the simplicity of its concept, its technical elegance, and its range of applicability carries the unmistakable Lang stamp, was employed for a whole series of proteins with excellent results.

These protein studies attracted again, as was to be expected, a large number of Danish and foreign guests, who contributed greatly to the production of the laboratory. In these last years Lang increasingly assumed the role of guiding and inspiring leader and counsellor, and although he could no longer take an active part personally in all the investigations of his colleagues, they nevertheless bore his stamp. It is impossible here to deal in detail with all these studies; one particular paper, however, must be mentioned. This is the long review, »Protein structure and enzyme activity,« published in 1959 by Lang and John A. Schellman. It is a masterly piece of work which recapitulates most of the results and ideas reaped during the last years of Lang's life, and relates them to the contributions of other workers in the same field. This summation is all the more imposing in demonstrating the progress which had been made since the 1951 Lane Medical Lectures; it also points up with painful clarity the great loss we suffered at the cessation of Lang's activities.

In the above I have attempted a more or less chronological review of the distinctive features of Lang's scientific and personal development. There are many sidelines to Lang's activities which have not been touched upon, and many collaborators who could not be mentioned. I will now conclude with some personal impressions and evaluations, and deal with certain aspects of Lang's life which I have not yet had occasion to mention.

It will undoubtedly be evident from the foregoing that the most characteristic feature of Lang's personality was the multiplicity of his talents. It is correct, I think, to designate him as a

25 Posthumous portrait of Linderstrøm-Lang.

genius. When one came in contact with him, one felt immediately the presence of an outstanding personality, a man whom one could not help admiring. I don't think there was anyone who denied him respect, and there were only very few who denied him confidence.

His talent for entertaining was a fable in all lands whose language he spoke well enough to express himself. Once, when he was scheduled to participate in a congress in the

United States, he was surprised to find that on the printed program for the final banquet, which he received prior to his departure, there stood simply, »In charge: Linderstrøm-Lang«. Without having the slightest idea of what was expected of him, he bravely set off for America. Needless to say, that banquet was a memorable one, during which he, among other things, delivered his famous lecture on »The thermodynamic activity of the male housefly«.

Lang's musicality was highly developed and genuine; the violin meant much to him, especially in his younger years. But he seldom went to concerts, preferring to make music in his own home. At the parties on Gl. Carlsbergvej both chamber music and Lang's singing played important roles. Shortly after the war, a British scientist came to Copenhagen, rather dignified and weighed down with his high official mission, and as Lang was having a dinner party that evening, he was invited to attend. The Englishman was hesitant, saying that he really had another engagement, but that he would be glad to come if he might leave just after dinner. During the evening Lang sang, told many of his innumerable stories, and finally, at 2 a.m., suddenly remembered his guest's reservation; he at once went over to him to ask about the other engagement. »Oh, forget about it,« was the answer,« I didn't know it was going to be *this* kind of party!«

Lang's »relapses«, as he called them, to painting were a somewhat serious matter in that while they lasted, he was completely obsessed and oblivious to anything else. These periods usually lasted a few months, with intervals of many years. The paintings were mostly used to decorate his home. He liked to experiment with different styles which resulted in pictures of a somewhat uneven quality, but I do remember that both C. O. Rude and V. H. Lundstrøm, two well-known Danish painters, expressed their sincere approval of the best of them. The most spontaneous expression of Lang's talent as a painter was in his caricatures in colored chalk of the staff members and guests of

26 One of Linderstrøm-Lang's many talents (1953).

the laboratory. These served as decorations at laboratory parties, and some of them are real gems.

In my own opinion, Lang's artistic talents were most perfectly expressed in his writings. His style had a peculiar warmth and directness which appeared spontaneous, but which was actually hard-won. Much labour was involved in wresting from himself those seemingly effortless phrases. As an essayist Lang was chiefly concerned with the peculiar position of the scientist in modern society – a problem whose scope he considered also to include epistemology. All forms of mysticism he treated as abomination and he was a master at ridiculing unnecessary verbiage by the aid of an anecdote or a revealing citation.

Lang never wrote a book, and although he sometimes considered the possibility of writing one on protein chemistry, I do not really believe he ever meant it seriously. His ability was not in this field, and he knew it. Lang was a man of ideas, and was bored by the labour inherent in compilation. He did write a number of review articles, but characteristically they all contain some original contribution to the subject, and some of his most fruitful ideas first appeared in these reviews.

Lang's position in Danish chemistry was naturally characterized in later years by his biochemical contributions. He was very much interested in Denmark's participation in the International Union of Biochemistry, in whose founding during the early fifties he himself had played a prominent role. In 1958 Lang was elected president of this Union. He was happy to retain contact with the Technical University through the Academy of Technical Sciences, of which he was president for a number of years.

Apart from science, Lang's activities were determined by two prominent traits of character: he possessed a high degree of social consciousness, and he could never say »no«. This naturally resulted in a long series of tasks and duties, nearly all of which were a burden for him. He acquitted himself very well on committees and attained various important positions, but this was due to his intelligence and conscientiousness rather than any interest in committee work *per se.*

Lang never really learned to enjoy a holiday. It is true that he was very fond of his sister's home in Jutland, his childhood summer refuge, to which he returned every year with his family; but the summer vacations seemed interminable to him, especially after he had got through the first dozen detective stories, and each year he reappeared in the laboratory just a few days earlier than he had originally planned.

Lang's numerous journeys were also characterized by his aversion to loafing. They were congress and lecture trips, usually comprising a whole series of different cities, and always loaded with meetings, official and informal, lectures, and entertainment. He loved to meet his colleagues in this way, and in

27 Linderstrøm-Lang and his wife Gerda in gay mood, facing their old friend David Glick (1949).

a quite unexplainable manner he thrived on this wearisome life, absorbing from it both knowledge and inspiration. Since he had little talent for the daily systematic perusal of scientific literature, his travels were to him an important source of information.

Official honors and distinctions were numerous, but they

were as nothing compared to Lang's personal popularity and the devotion and affection which he was able to elicit from widely different types of people. He was himself deeply and truly interested in his fellow men, and there was no limit to his readiness and willingness to help others. He loved to give, and his lack of closefistedness, both in money matters and in human service, was complete. Very often he became a father confessor, even for persons who were not close friends, and his gentleness never failed.

It is obvious that many problems must arise for the wife of a man who was devoted to so many things. As one who has known the Lang household for many years, I may perhaps permit myself to say that Fru Gerda Lang has truly mastered the difficult art of being the wife of a great man. Her hospitality was unlimited, just as his, and she was the most natural and unassuming hostess imaginable. She accompanied Lang on many of his travels and was always an unfailing support to him.

In his younger years Lang liked to play tennis, but he was no outdoor man and for long periods his desire for exercise was satisfied with peripatetic wanderings in the laboratory. He suffered much from migraine, but otherwise was by no means in poor health – on the contrary, he could tax his constitution with much hard work and irregular habits without ill effects. When he was 50, he contracted diabetes, but in a mild form which caused him little discomfort; it was, however, to have fateful consequences in his final illness. In March 1958 he was operated on for a relatively benign tumour of the intestine, which was removed completely. But following the operation, typical diabetic complications appeared, and after a long year in hospital he died on the 25th of May, 1959.

Danish science has in Lang lost one of its most prominent figures, prominent by virtue of his scientific qualifications and his personality. He was appreciated by his countrymen, and he repaid this feeling with a deep love of Denmark, its landscape, its people, language, and customs. His stature and integrity were as great as his work, and throughout the world he will be missed by all who knew him.

It has been stated in the preceding biography that Linderstrøm-Lang could write beautiful prose. He also was a man deeply interested in the social responsibilities of the scientist, a theme that at the time of his life was by no means as generally debated as it is now. He discussed this theme in a number of essays and lectures and was especially concerned with the fact that Denmark was rather tardy in setting up the research councils which he regarded as a necessity. As a sample for his concern about the future of mankind, and as a sample of his prose, we reprint here a lecture which Linderstrøm-Lang gave in 1957, as the opening address to a congress of the European Brewery Convention in Copenhagen. In giving this lecture, Linderstrøm-Lang followed a tradition from the time of S. P. L. Sørensen, who repeatedly addressed similar audiences. But while Sørensen confined himself to the natural sciences and their impact on technology, Linderstrøm-Lang took a much broader, humanitarian view.

Man, science, and industrial development

by *K. Linderstrøm-Lang**

The driving force in the development of human scientific accomplishments is essentially dual. In the old societies I find it symbolized in the meditation of priests upon the mysteries of nature and in the toiling of craftsmen and labourers with the conquest of matter. This dualism arose originally from the deep chasm between human understanding of the universe and the ability of man to live in it. At the same time there was and still is a close connection between these two aspects of human relation to nature: without papyrus, clay tablets, metals, glass, cyclotrons no fertile meditation or scientific synthesis is possible. On the other hand only a slow and laborious conquest of nature can be accomplished without perceptive synthesis. The influence of astronomy and surveying on agricultural development in ancient Babylon and Egypt and the importance of modern theoretical physics to the development of atomic energy may serve as examples. It is well-known that in his pursuit of applied science the labourer – forgive this simplification – has had the less enviable position in the known historical societies. The structure of these societies was such that the blessings of labour were bestowed upon the majority of their citizens, while the burden of leisure was laid upon the shoulders of a minority – who were suitably indemnified through their better living conditions. Idleness is not always the manure of the evil one. In proper minds it may give rise to thoughts which force open the entrance to the secrets of the universe. We therefore rightly praise the great scientific per-

**Written in 1957.*

sonalities and their schools who have maintained the integrity of pure science over the centuries, but we do not properly appreciate the everyday heroes of the past and present, women and men who, amidst slavery and indigence, through luck or deliberate experimentation, have created the material basis for these cultures.

In the course of time a hybridization has taken place between the worker and the priest, but at a greatly different rate in different branches of science. In medicine for example the dualism disappeared already in ancient Greece. Nor did this process follow a smoothly rising curve. Schools have arisen, flourished and died. Universities have struggled for freedom, gained freedom and lost freedom in conflicts or collaboration with governments, churches or finance. But the process went on. Craftsmanship turned into industry and science. Tycho Brahe and Galilei built their own astronomical instruments, Van Leeuwenhoek watched microcosms through self-made lenses, alchemists ruined princes by their experiments, Pasteur came to the rescue of the viticulture of France, F. Haber was the founder of the modern nitrogen industry, Merck & Co., Inc. synthesized folic acid. And let us not forget the surgeons: master artisans and scientists.

This interaction between speculation and production has tended to reduce the distance between experiment and theory, so that in our time the masters in pure science are not placed in a more favourable light than their brothers in applied science, a state of affairs which is intimately connected with the growth and structure of human societies of today.

I have ventured to give this crude exposition because people in this science-ridden century still believe that scientists are particularly free in their actions, and that, because of the dangerous consequences of these actions, they are a peril to humanity. It has often been said that the growth of knowledge is not worth the human sufferings it may cause.

The fact is that there is very little sense in this talk about the freedom of scientists. As I have tried to intimate, scientists are no longer detached gods – if ever they were. Possibly the students of particular sciences like pure mathematics are relatively free, but that is only because these sciences are regarded as harmless or useless to the state, the church, or finance. In our time it is especially the question of financial support to science which is important. Science has become a costly affair, its expenses far exceed what even wealthy individuals can afford, and the state or finance must assist. In other words: the scientists are citizens of society for better or worse.

As far as I can see the reason the tree of knowledge is a gift of good and evil must be sought in those sides of human nature which have been subjected belatedly and unsatisfactorily to rational investigation with scientific methods. The increase in our understanding of nature is in itself undoubtedly progress, but as soon as we come to the inclusion of this knowledge in human societies as an element of technical power, progress is a much more ambiguous word. It is quite evident that this process of inclusion does not proceed – and has never proceeded – in a harmonious way, but when we consider the unbounded disregard of harmony which characterizes living nature, it is hardly to be expected that the ascent of mankind to a greater command of nature would be a peaceful one. Nevertheless the question frequently arises whether science, which has bestowed so many blessings on mankind, could not also teach mankind to enjoy these blessings in concord and peace.

I have often met the opinion that if only a fraction of the costs of science had been spent on the study of man and man's relation to society, the world would have been a better one today. This is to cast an unreasonable reproach upon science. What is it in fact which has cost billions – thousands of years of misery, blood, sweat, and tears, – if it is not the study of man? What have poets, painters, and philosophers accomplished, if they have not studied and taught man? What have historians, national economists, philologists, archeologists, ethnographers, and psychiatrists done? What, finally, have the great lawmakers and reformers of society done, if – for good or evil – they have not studied man and experimented with him? It is possible that these efforts have given poor results, but it is certainly not because of an insufficient expense account. The problem has simply resisted analysis by means of the principles which have been so fruitful in science, and if that is the case intellectual unity among men is impossible, the words »true« and »untrue« become irrelevant, »right« and »wrong« take their places.

There is an essential feature in the study of man and society which I should like to elaborate upon. The word »law« may be used in several senses. When we talk about scientific laws we know what we mean.

But the concept »laws of societies« is ambiguous in that it includes both rules of conduct and laws in a more »scientific« sense. Still the rules of conduct must in some way be connected with the scientific laws of society. They must be derived from them and incorporated into them. The ten commandments are such rules of conduct based upon laboriously acquired experience and upon an internal need of stabilization in the ancient societies. However they have been disobeyed, their importance for the development of our societies cannot be overrated. The declaration of the rights of man, which Jefferson so beautifully interpreted in his speeches, is the promise of society to its citizens; but at the same time it establishes rules of conduct derived from experience. The Communist Manifesto is a similar conglomeration of promises, rules of conduct, and laws.

I shall not attempt to weigh these manifestations of human laws against one another, but just point to the fundamental interaction occuring here between subject and object, between the experimenting observer and the object of his observation.

The problem is not new in psychology and biology and it has

recently been met with in quantum physics where – essentially through the inspiration of N. Bohr – it has led to a revolution within epistemology and has influenced science as a whole.

I am sure it is well-known to you that the quantum theory has set a limit to the prediction of physical events on the level of the single electron and that physical laws in our time express probabilities of events, or statistical averages over large numbers of elementary particles. Drawing the parallel from the behaviour of electrons to that of individuals in a society it seems tempting to suggest that while the future of an individual is unpredictable, that of society as a whole can be described on a statistical basis. In this way both the individualist and the social scientists may be accommodated.

However, it is hardly necessary to emphasize the extreme looseness of this analogy. I would not like to be in the shoes of the social scientists when it comes to deriving the laws of society from history and social experiments. I can only express the hope that when a solution has been found it has not cost too many tears and too much blood. Sweating does no harm.

What is now the position of natural science in this matter? In the colossal drama in which the laws of societies are created – and are observed and described while they are being created (whereby they are changed) – we are both audience and actors. Far from securing success – as one might imagine – this arrangement gives rise to occasional or continuous brawls which draw their bitterness from the fact that nobody can leave the theatre – at least temporarily. Cast and producer are under violent cross-fire, the police and the fire department must often intervene. In this theatre the scientist is the handy-man. He looks after the heating of the premises and the upholstery of the seats. He conditions the air and supplies the bar with liquor. He tends those who have fainted. He is the stage manager and fires the appropriate shot at the climax of the drama. Sometimes he even writes the play. But in addition he often acts the part of »the man who talks reason« – a faint voice most certainly, but one which is heard politely in our time. After all the fellow is useful in many ways.

As a matter of fact, to talk reason ought to be natural for a scientist. Not because scientists are more reasonable by birth than other people, but because science is reasonable, science is rational. I really think that it should have a mission here. In order to show what I mean, permit me to tell a story about my most beloved and admired teacher and predecessor at Carlsberg, S. P. L. Sørensen. Sometime in his middle-age he took part in a spiritualistic session. Where and why has escaped my memory. But there he sat at a table together with a group of others – quite distinguished people, so he told me, and most of them disbelievers like himself. In the course of the evening the circle succeeded in producing several uncanny raps while a weird and ghostly atmosphere pervaded the room. One of the participants left the table and was set to guard a purse which contained an unknown sum of money. The spirits were then supposed to tell the magnitude of the sum – only the number of crowns, I think, because after all the means of communication, the table-legs, were imperfect. The table rapped 9 times, and sure enough there were 9 crowns in the purse, neither more nor less. The light was dim, the bookshelves creaked, the wind whined in the chimney, and the people present were more or less prepared beforehand through eery stories of ghosts in other sitting rooms. I was therefore inclined to believe Sørensen, when he told me that everybody got pale around the gills and wanted to go home to bed after this first encounter with the transcendental. But Sørensen said the divine words: »Now we do this 10 times«, and actually forced the participants to repeat the experiment again and again. The result was entirely negative – which of course may have been luck in the opposite direction. But that is not the point of the story. Its moral, as I see it, is the victory of scientific reasoning and method over emotional evaluation. In everyday life we are all too frequently satisfied with our own judgement based upon entirely insufficient observations and coloured by our personal fancies. In a wider connection it is the appeal to rationalism and even to tolerance which I find so stimulating in Sørensen's story.

During the occupation of Denmark I gave a speech in 1942 at Aarhus University and told the following story about two boys, Alfred and Emil. Today I would call them Adolf and Moritz, but let us not change nomenclature: Alfred is sitting in the class room and next to him is Emil who has the habit of always putting up his hand just a few seconds before Alfred. He has other characteristics which Alfred dislikes without seeing very clearly that the dislike assumes unreasonable dimensions *because* Emil puts up his hand too early. Let us say that Alfred really is a nice boy but an introvert type, inclined to brood on grievances. He may possibly be less bright than Emil, or he may think more slowly or dread making a fool of himself. Emil, on his side, may bluff a little and put up his hand before he really knows what to answer. Emil may be able to think while he is being questioned; Alfred gets stuck. Possibly Emil is the favourite of the teacher, possibly he is a bad egg. One thing or another, Alfred gets mad. He broods about Emil, the problem Emil, he becomes possessed by the thought that it is Emil's fault that he – Alfred – does not get along in school so well as his mother thinks he deserves. He begins to see the worst in Emil's character. If Emil is friendly, he is obsequious; if he is frank, he is brazen, if he is reserved, he is haughty. A new Emil is created which may – or may not – be like the real Emil. Alfred wonders why Emil is such a beast. There is of course his appearance, which is bad, almost disgusting. He does not resemble the other boys, or it may only be Alfred's Emil who is different. Anyhow, Alfred senses a sinister connection, he gets suspicious of people who resemble Emil, and when one day he meets Emil's father – who resembles Emil and is obsequious when he should be friendly, brazen, when he should be frank – the scales fall from Alfred's eyes, and he realizes all of a sudden that Emil owes his detestable properties to his father. It is in the family!

With this discovery Alfred has started to generalize. Before, there was just a tragic affair between two boys, Alfred and Emil. But now a quasi-scientific question has arisen about inheritance of qualities of the mind along with physical appearance. How far this avalanche may roll, if we assume that Alfred is inclined to further generalizations, should not be difficult to imagine.

It is along such dubious routes that all too frequently we arrive at our convictions about the qualities of men, races and nations.

So much about science and tolerance.

As regards the practical importance of science in everyday life and during festivals, its contribution to technical production and comfort, to the mitigation of illness and the destruction of innocents, one can never be wrong in stating that it is immense. However, the question about its absolute value, beneficial or pernicious, is undoubtedly irrelevant. There is no answer. A. Huxley, who has talked wisely about the curse of the technical development, its favouring of oligarchy, »uniformation« and slavery, proposes decentralization and decrease of production as the salvation for humanity, but in view of the general unanimity of the world about quite opposite measures the idea is impractical. Actually there is just as much wisdom in talking about the horrible social crises, the abyss of barbarism which a decrease of production would create in the overcrowded world. After all we prefer the radio to the wooden horse.

As far as I can see the only thing we kan do is to say: »Now, here we have a tool, science. Whether the world remains split up in different societies and cultures or whether it becomes united into one, this tool, the product of the instinct of selfpreservation and the human brain, is a means of making us esteemed among citizens of the world and of providing us with food, light, clothes and a certain measure of human freedom.

Hence let us use this tool as well and as rationally as we possibly can, and try to prevent it from doing what we believe is harm, where it was meant to do what we consider good.«

How can such a matter be approached, how can scientific thought be peacefully incorporated in the minds of people? How can scientific methods and results be made useful to everybody, how can ruthless exploitation of the waning resources of the earth be fought? There is no doubt that we live in an era in which governments exert a steadily increasing control on production. Whether this trend is temporary or not remains to be seen. I do not consider it a passing phenomenon and have observed with considerable satisfaction that science is building up organizations which are not without power to assist governments and other bodies in the difficult task of rationalizing the activity of basic and technical research for the benefit of mankind as a whole. I refer here to the research councils which are set up in many countries and serve to promote science in all its aspects. These research councils are composed of scientists and officers of the government and they are secured considerable independence in matters concerning organization and financing of science.

I think it is appropriate on this occasion, speaking to an audience of distinguished members of the Brewers' Guild, that I mention the »Research Council« set up in Denmark many years ago, not by the government, but by a far-sighted and idealistic industrial leader, J. C. Jacobsen. In his deed of gift to the Royal Danish Academy of Sciences and Letters, concerning the establishment of the Carlsberg Foundation, Jacobsen wrote: »Since an institute for special research (like the Carlsberg Laboratory) can only thrive if it is uplifted by the spirit and penetrated by the light which radiates from the sciences as a whole, and since this light has ever been a source of bliss and comfort to me, it is my heart-felt desire, as an instalment of my debt, to partake in the support of science in general, and particularly of those fields where the state has failed in the past –

and probably will fail in the future – to provide the necessary funds.«

»I here allude to purposes such as payment of subventions to younger scientists, salaries to persons who have distinguished themselves particularly in science, travelling stipends, and, in general, grants in support of scientific investigations of particular and immediate importance, etc.«

»In order that an institution of this kind be able to serve its purpose now and in the future, it must be ruled by an executive committee, elected among men of scientific insight and ability« – and Jacobsen here goes on to explain why he has turned to the Academy of Sciences and Letters with his request that the committee be elected among its members. He emphasized especially the integrity of the Academy in a political sense as a reason for his choice. I cannot but see that we have here an anticipation of the research councils of our time, and the encouraging results of the activity of the Carlsberg Foundation through 80 years emphasize the great importance of such institutions to the scientific life of a country.

However, the »scientification« of society, the incorporation of scientific thought in the minds of people and of scientific methods and results in the machinery of society is necessarily a long drawn out process. Let me just mention the problem of recruitment of scientists. It has greatly alarmed the organizers of scientific technology, particularly those engaged in the development of atomic power. What is the sense in pulling scientists in as members of our research councils, consultants to our institutions, workers in our industries, if there is not a sufficient number of properly educated people who are willing to work? I admit that there is a terrible shortage of skilled labour on a high educational level – all over the world – but as long as most children, everywhere, take their bicycles apart, fish for tadpoles, and burn holes in their mother's carpets with chemicals, there is still hope. To those who fear an overproduction of intelligence, I should like to say that it is a danger we should meet with composure.

Naturally there is hard work to be done, spreading the love and knowledge of science among all members of society, catching souls – children through parents and vice versa. In our time the situation is the unhappy one that costly and complex instruments have become inserted between object and observer and that the language, or the dialect, if you like, in which the observations are described has a foreign accent to the layman. The problem here is to find a presentation which is equally far from sensation and from triviality. I have always compared the eminent writers of popular science with the master-chef who can make a hare taste more of hare than intended by Providence – and more digestible. But such masters are rare.

Then there is also the widely spread horror of organization to overcome, the devotion to ancestral ideas about the »display of free forces«, a relic from the time when the earth was wide and defenceless against the white man. I shall admit that by birth and upbringing I belong to this period, but I regard my fear of organization as a sort of claustrophobia which must be fought down with arguments. It must be reassuring to the scientists that they are themselves in charge of this organization.

Finally international relations should be mentioned. It is too early to evaluate the results – on a political level – of the unselfish work of Unesco and the numerous international scientific unions, but to me it is one of the brighter sides of an uneasy and disunited world. Speaking to this audience who represent an international organization of scientists and technologists, where the scientists are technologists, and the technologists scientists – is, as Jacobsen said, a source of bliss and comfort to me. The principles of collaboration and open exchange of ideas which you advocate is actually uplifted by the spirit and penetrated by the light which radiates from science as a whole.

Carsten Olsen

1891–1974

by *Poul Larsen*

Carsten Erik Olsen was born March 1, 1891 in Copenhagen as son of a post-master. When Carsten was 11 years old, the family moved to Thisted in Northern Jutland. They lived in Thisted for 2½ years, and there is no doubt that this relatively short period has been of decisive importance for Olsen's interest in natural history. He wandered around in the Limfjord district and began collecting insects and plants. In 1904 the family moved back to Copenhagen, where Olsen finished high school in 1910. Then he studied natural history at the University of Copenhagen and took his final degree in botany in 1916.

During his university studies and through the following years Olsen was mainly occupied by vegetation analyses in Danish plant communities as well as their ecology. This interest he preserved throughout his life, and he kept it up by excursions, mainly in Gribskov forest north of Copenhagen, where his summer cottage was situated in the small village Gadevang.

In 1914, while he was still a student, he published some fairly extensive papers. One concerned the family Cornaceae in the series »The structure and biology of arctic flowering plants«, another the vegetation of some Sphagnum bogs in North Sealand. His interest in mosses continued: in 1917 he published an investigation of the immigration pattern of epiphytic mosses on the bark of various trees, and in 1917, 1920 and 1946 he described the changes in the moss flora of a bog (»Maglemose«) in Gribskov forest. Once in a while he included species of Sphagnum in his experiments on plant nutrition.

As a student, Carsten Olsen was influenced by the two great

28 Carsten Olsen in 1940.

Danish ecologists, E. Warming and C. Raunkiær, especially by the latter whose techniques and ideas he mastered thoroughly. In 1918 Olsen was awarded the University's gold medal for a prize essay »A morphological, biological and statistical investigation of the mosses and vascular plants in the ground flora of Danish oak forests and oak scrubs«. At the time he received the prize, Olsen was already deeply involved in his investigations of the relations between plant growth and the hydrogen ion concentration of the soil, and therefore the results of the oak forest work were not published till 1938 after considerable pressure from several colleagues including Raunkiær himself.

Warming and Raunkiær, the ecologists, have on various occasions stated that true ecological work always ends up in the laboratory. Olsen conformed to this rule. His investigations of the influence of hydrogen ion concentration upon plant growth were begun in 1916 (while he still was busy with his oak forest work), and the experimental part of this research was performed at the Carlsberg Laboratory, where S. P. L. Sørensen, the hydrogen ion specialist and father of the term pH, took great interest in Olsen's work.

The task that Olsen set himself was to get to the bottom of the complex of problems set by the calcicole and calcifuge plants: Why are some species found only on calcareous soil, others never? To seek the main reason in the difference in hydrogen ion concentration was only one – and a relatively new one – of the possibilities that had been discussed. Olsen decided to tackle the question by combining Sørensen's methods of pH-measurement with the formation-statistical techniques of Raunkiær and set to work.

He investigated 76 localities on grassland and 188 localities in woods. Raunkiær's formation-statistical method (populary called »circling«) involves throwing a steel wire circle of 0.1 m^2 area at random in a number of spots in the plant formation to be investigated. All species in the encircled areas are noted down. That is to say one must be able to identify all the species

29 Olsen performing an experiment on pH variation in plant water cultures (1917).

present, in all stages of development. Olsen covered 264 localities, each comprising a great number of 0.1 m^2 circles. So the field work alone represented a vast amount of labour, to say nothing of the pH-measurements in the laboratory.

Olsen was able to show that a great many species are to be found only on soils within a certain range of pH, characteristic of each species. He also confirmed his findings in nature by means of water culture experiments in the laboratory. But although the relation between hydrogen ion concentration and plant distribution in nature thus was convincingly documented, Olsen himself fully realized that his results did not clear up the mechanism responsible for the relations found. The effects of the hydrogen ion concentration on plants might very well be indirect, e.g. due to variations in soil constituents, which in turn were dependent on pH. In fact, most of Olsen's subsequent scientific production had its roots in his desire to solve the problems here encountered.

1921 Olsen published his results in his doctor's thesis: »Studies on the hydrogen ion concentration of the soil and its

30 Water culture experiment with wavy hair-grass (Deschampsia flexuosa). Left: pH 4, iron source ferric sulphate. Center: pH 7, same iron source. Right: pH 7, iron versenate. This picture gives a good example of Olsen's skill in presenting convincing evidence.

significance to vegetation, especially the natural distribution of plants«. This work quickly made him famous. It brought order and reason to a field of ecological plant physiology which until then had been very controversial and confused. To this day, it is frequently quoted in the relevant literature, and it has rightly been called a »classic«.

An important subject in Olsen's later work is the absorption of iron by plants. Deschampsia flexuosa (Wavy Hair-grass) for instance, a plant that is only found on acid soils (pH 3.5-5) can be made to grow beautifully in water culture even at pH 7 and 8, if iron is given not in the normal form of ferric chloride but in a complex compound (iron versenate) which is not precipitated at high pH-values. This means that Deschampsia is confined to acid soil not because it is damaged by high concentra-

31 Olsen in his laboratory (1936).

tions of hydroxyl ion but because it cannot take up enough iron from alkaline soils. Similarly, Helodea canadensis (Pondweed) depends strongly on the availability of iron as ferrous iron or in the form of soluble, complex compounds. This is the reason for this plant's occurrence as a pest in ponds with bottoms poor in oxygen, where iron occurs as ferrous sulphide. In culture, if iron is available only as ferric chloride or ferric citrate, Helodea does not thrive. Iron versenate, on the other hand, is an excellent source of iron.

An investigation along similar lines concerns the question of why beech cannot grow at certain patches in Allindelille forest, some of the most alkaline localities (soil pH 8.2) that Olsen had encountered in his thesis work of 1920, and to which he returned in 1943. He could now show that the reason for the absence of beech was not the local overabundance of calcium carbonate but the fact that certain substances, especially iron and nitrogen were not present in the alkaline soil in a form available to the plants.

The uptake of nitrogen by plants had interested Olsen ever since 1920. This led to studies of nitrate formation (already in his doctor's thesis) and to studies of nitrogen fixation in bacteria and in blue-green algae, especially a species of algae living in symbiosis with the water fern Azolla. In 1950 Olsen dumbfounded the local florists by introducing Azolla into a small pond in the vicinity of his summer cottage with the purpose of combating duckweed. Azolla covered the pond with a solid, brick-red carpet (Olsen later showed that Azolla, thanks to its symbiont can fix 95 kg nitrogen per hectare during a summer). Then in the winter, the cold killed the tropical fern, restoring the pond but with a greatly reduced population of duckweed. Olsen, as a matter of form, reported the experiment to the botanical authorities and according to Lecturer Erik Bille Hansen, got the answer: »So *you* were the culprit!«

Carsten Olsen began his career in the Carlsberg Laboratory in 1916 as a visitor to the Chemical Department. He was appointed assistant in the department in 1919 and held this position until 1944. At that time he was promoted to a more independent position, that of »Amanuensis extraordinary«, directly under the board of trustees. He remained in this position until 1961, when he retired at the age of 70. He had at that time served the laboratory for 45 years, longer than any other associate of the laboratory.

Carsten Olsen was accustomed to work alone. He produced his own problems, his own solutions, and his own techniques. One of his very few collaborators was the physicist George Hevesy who had considered the idea of using radioactive isotopes as tracers in biological research (for which he received the Nobel prize in 1943), and who in the 1940s turned to Olsen for help in the solution of some plant physiological problems involving the use of radioactive phosphorus. But Olsen disli-

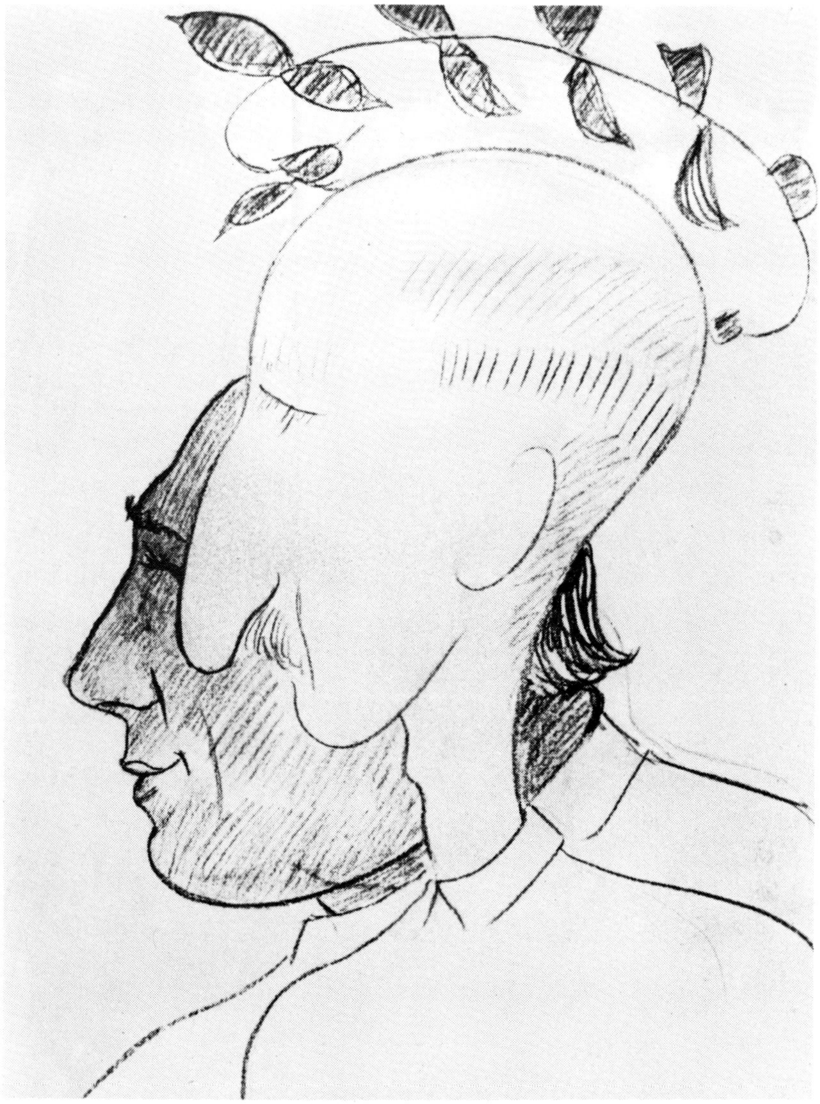

32 Linderstrøm-Lang's Cervantesque caricature of C. Olsen and G. Hevesy, who were pioneers in the introduction of radioactive isotopes in plant biochemistry (1937).

ked to be dependent on somebody else's ideas and experimental skills; he always regarded this work as a sideline and was glad to return to his own problems when the German occupation forced Hevesy to leave Denmark.

As a result of his lonely habits, Olsen had no pupils. But indirectly, by the impact of his work, he had a profound influence on Danish plant physiology. As a lecturer, he was much appreciated, especially in the Danish Botanical Society and the Scandinavian Society for Physiological Botany. In 1951 he was elected member of the Royal Danish Academy of Sciences and Letters. He participated in a Symposium in Southampton in 1960; but otherwise he travelled little, preferring the world of the Danish forests, his laboratory and his greenhouse. These occupied him to the last. He published his last paper in 1971 at the age of 80. Three years later, on August 19, 1974 his industrious life came to an end. He left his wife, née Elsa Eriksson, whom he had married in 1926, and a son.

Through his work Carsten Olsen has contributed fundamentally to the clearing up of many controversial problems regarding the mineral and nitrogen nutrition of plants and their dependence on the soil, especially its hydrogen ion concentration. His works are characterized by his severely critical turn of mind in the experimental techniques and by an impressive lucidity in the planning of an investigation and in reporting its results.

Mogens Westergaard
1912–1975

by *D. von Wettstein*

Mogens Westergaard passed away on June 8, 1975. To that very day he was ardently probing his ideas about chromosome pairing and crossing-over on his latest pet, the meiosis of lily. He loved to nurture the plants, to produce the slides by trying various techniques, to observe the chromosomes and to discover phenomena neglected or misinterpreted by previous investigators. Prepared by a thorough review of the problem to be analysed, he was a master in using the cytology, which gave him so much satisfaction, as a tool to solve major genetic problems. Deteriorating health over the last years made it increasingly more difficult for him to pursue his goals as energetically as he would have liked to. With the courageous and generous support of his trusted wife Ebba, the physical handicap was conquered again and again. Even very recently, Mogens Westergaard presented us with a most challenging preselection hypothesis of crossing-over.

Westergaard was born on June 12, 1912 in Copenhagen. As a boy he was of rather delicate health. Especially a case of bone tuberculosis was a serious problem, but was successfully overcome by surgery which, however, left facial scars visible throughout his life. He finished high school in Horsens (Jutland) in 1931 and went from there to the University of Copenhagen, where he took his degree in botany in 1936. In 1940 he took his doctor's degree with the thesis on *Melandrium album* which is more thoroughly discussed on later pages.

Most of the work on the thesis was done at the Carlsberg Laboratory in Professor Ø. Winge's Physiological Department, but he was at the same time (1936–1949) assistant in the Gene-

33 Mogens Westergaard (1955).

tical Laboratory of the Veterinary and Agricultural University. There was no chair of genetics at the University of Copenhagen at the time and after the war, in 1945, Westergaard was appointed as a lecturer at the university with the task of teaching the genetics course. During the following years, he fought a heroic battle with the authorities (which were harassed by the financial difficulties of the post-war period) in order to establish the Genetic Institute which the university had sorely needed for many years. He finally succeeded and was in 1949 appointed the first professor of general genetics at the university of Copenhagen. In 1961, he retired from the chair of genetics and returned to the Carlsberg Laboratory, where his scientific career had begun. Mogens Westergaard took great interest in the welfare of the Carlsberg Laboratory and represented an initiating and decisive voice in the development which now has provided the laboratory with the necessary new facilities for its continued scientific activity. He felt that traditions are very valuable but not sufficient to support a laboratory. He and A. Tovborg Jensen were able to win the executive director of the Carlsberg Breweries A. W. Nielsen to cherish the Carlsberg Laboratory as a basic research institution. The industrial manager and the scientist also shared the view that Danish industry had to amplify its applied research efforts. Poul Brandt Rehberg and A. W. Nielsen found the necessary ways of reorganisation, which allowed the breweries' extended laboratories for research and development to move together with the improved facilities of the Carlsberg Laboratory on Gamle Carlsbergvej. A new dimension was thereby added to brewer Jacobsen's intentions according to which basic and applied research in the natural sciences should form the platform for industrial development.

In the intellectual pedigrees drawn by A. H. Sturtevant for his history of genetics (1965), Mogens Westergaard descends in direct line from Wilhelm Johannsen and Ø. Winge, the latter in fact being his mentor together with C. A. Jørgensen. Westergaard's first outstanding contribution to genetics was the discovery that the male sex in the dioecious white campion, *Melandrium album* is determined by the presence of the Y-chromosome. His clarification (in the years 1936 to 1948) of this genetic mechanism of sex determination with the aid of polyploids and fragmented Y-chromosomes is a beautiful synthesis of the lines of experimentation developed in Copenhagen by Winge and Jørgensen. The results represent an important cornerstone of cytogenetics and paved the way for recognition of this mechanism of sex determination as the one valid for mouse and man.

Mogens Westergaard was one of the first to recognize that chemical mutagenesis constitutes an important instrument for the molecular analysis of the gene. He developed the back mutation test for an adenine requiring strain in the fungus *Neurospora* and found that the back mutation pattern of a given mutant gene depends upon how the gene was originally damaged. This led to the concept that the back mutation pattern can provide information about the nature of the original forward mutation, a concept exploited successfully by E. Freese, S. Benzer and others in the establishment of changes, insertions, or deletions of one or a few nucleotides as the molecular mechanisms of gene mutation. Westergaard's contributions to chemical mutagenesis span the years 1947 to 1960.

In his analysis of meiotic chromosome pairing and crossing-over, he chose to work with the ascomycete *Neottiella,* a fungus that still cannot be cultured but which is a superior material for cytochemical studies and the ultrastructural analysis of meiosis. He definitively laid to rest the copy-choice theory of crossing-over by showing that meiotic DNA replication takes place before the two haploid nuclei fuse, i.e., before the homologous chromosomes are within the same nucleus. The structure of the leptotene chromosome was clarified and it was discovered that the chiasmata at early diplotene consist of short pieces of the synaptinemal complex which hold the repulsing chromosomes together. This led to the conclusion that the molecular pairing of homologous DNA segments of two non-sister chromatids and crossing-over takes place inside the synaptinemal complex, the special ribonucleoprotein structure which is found between the paired homologous chromosomes of a pachytene bivalent.

Apart from the mentioned three areas of research, which highlight Mogens Westergaard's fundamental contributions to genetics, his scientific production mirrors two great hobbies: botanizing as a means of recreation, and visits to the beautiful places of the earth. The objects include a species of *Iris* from the little island of Saltholm, hybrids of *Calamagrostis* and *Ammophila,* cytotaxonomy of the flora in Greenland, desert plants in Death Valley, and eel populations from different oceans. Let us have a closer look at his main contributions to genetics: At the time when Westergaard started his studies of sex determination in *Melandrium*, it was generally believed that the ratio between the number of X-chromosomes and autosomes determines whether the individual is male or female. The sex deciding genes are located in the X-chromosome, whereas the Y-chromosome is without influence. Westergaard's results from triploid and tetraploid *Melandrium* plants as well as from different aneuploids reveal that the Y-chromosome is male determining and that it takes at least four X-chromosomes to balance the male determining genes in a single Y-chromosome. This conclusion was independently reached by H. E. Warmke.

Westergaard's classical paper of 1940 also contains the correct interpretation of chiasma formation for XY bi- and multivalents within the framework of C. D. Darlington's theory of sex bivalents as permanent structural hybrids. The X- and Y-chromosomes have a short terminal homologous segment and only there can chiasma formation take place. I remember the happiness and excitement with which he studied the three dimensional reconstructions of the ultrastructure of the sex bivalents made by A. Solari in the mouse, rat, and man. At the occasion of preparing a review on the synaptinemal complex for the Annual Review of Genetics (1972), he realized that these reconstructions provided for the first time a convincing cytological demonstration of the validity of the theory of sex bivalents as permanent structural hybrids. The short homologous segments of the X- and Y-chromosome form at pachytene a synaptinemal complex. At diplotene the latter is converted into a chiasma as a result of crossing-over, whereas the differential segments of the sex chromosomes remain unpaired due to the development of a special structure. In the differential segments are found – besides the sex deciding genes – those that show X-linked and Y-linked inheritance respectively.

In his paper of 1940 Westergaard was able to refute H. J. Muller's hypothesis which states that the balance between sex chromosomes and autosomes as a sex determining mechanism is responsible for the rare occurrence of polyploids in the animal kingdom. In *Melandrium* it was possible to establish tetraploid populations with a functioning mechanism of sex determination.

By analysing plants with deletions for parts of the Y-chromosome, Westergaard characterized the sex deciding genes that are located in the differential segment of the Y-chromosome. In the terminal end of the segment are located one or several suppressor genes for the development of the female sex organs. Removal of the chromosome part closest to the homologous segment results in male sterility while the centromere region of the differential segment contains genes that control anther formation. It thus emerges that the Y-chromosome in *Melandrium* determines the male sex by absolute linkage in its differential segment of suppressor genes for female development with genes functioning in the formation of male sex organs and gametes.

A major portion of the *Melandrium* studies were completed when the war interrupted Westergaard's scientific work and he joined the Danish resistance movement. With great personal courage, he participated in the military activities of the movement and shared eventually with many other members internment in the Frøslev camp. During this period, Westergaard became acquainted with a large number of persons, who like himself were to play a major role in the unprecedented development of research and higher education in Denmark that took place in the decades after the war. This confrontation with political encroachment and the experiences gained in resisting them allowed Westergaard to present in 1948 a most competent and understanding analysis of the development of the T. D. Lysenko affair, which by political backing of misconceptions on heredity led to the destruction of genetic research in the Sovjet Union.

In 1946 Mogens Westergaard went as a Rockefeller fellow via the John Innes Institute to the California Institute of Technology for a year. Here in the Mecca of the newly developed field of biochemical genetics, he worked out in cooperation with H. Mitchell the medium that facilitates fruiting body formation in *Neurospora,* and he conceived the back mutation test. (This test was independently worked out the same year by N. H. Giles.) His paper from 1957 entitled »Chemical mutagenesis in relation to the concept of the gene« summarizes the results obtained with this approach mainly in collaboration with Gunnar Kølmark. It had a profound influence on the thinking of many researchers in the field. The experimental results obtained by Westergaard, Giles and Kølmark with the back mutation test have helped us to recognize that the nucleotide is the unit of mutation and recombination, whereas the gene represents a unit of function.

In connection with or rather simultaneous with the chemical mutagenesis projects, Westergaard became a primus motor in the development of biochemical and molecular genetics at the national as well as international level. He created the Genetic Institute at Copenhagen University as a teaching and research laboratory which was frequented from the start by students and visiting scientists from many parts of the world. During the first years, the laboratory was located in premises borrowed from H. V. Brøndsted at the Institute of General Zoology. In an administrative ordeal he got built in the Botanical Garden the laboratory facilities, which had been conceived 25 years previously by Winge, C. H. Ostenfeld and P. Boysen Jensen. It was to be a happy marriage for students and teachers alike, to house besides genetics in the new building, microbiology, biochemistry and plant physiology. Aided by his student Ove Frydenberg – specializing in *Drosophila* and population genetics – he organized courses comprising classical as well as molecular genetics. As dean of the faculty, Mogens Westergaard succeeded in establishing truly modern study curricula for biology, curricula that are still in use basically unchanged. In order to breathe new life into European genetic research, he organized together with Boris Ephrussi, Guido Pontecorvo and David Catcheside six scientifically and socially superb symposia supported by the Rockefeller Foundation. They had the specific purpose of encouraging collaboration between European and American geneticists working on biochemical and molecular genetics. The list of such activities during his association with Copenhagen University is fragmentary, but sufficient to understand that the mind was asking more than a physically frail constitution could support. This period ended towards the end of 1961 in a near fatal coronary thrombosis.

He had started his cytological and genetic research as a student at the Carlsberg Laboratory, to where he now returned, the Carlsberg Foundation providing him with the facilities to continue his research. His studies on the mechanism of crossing-over have two roots. One was a »show down between genetics and mycology« at the International Botanical Congress in Amsterdam in 1935, where he travelled by bicycle from Copenhagen visiting Emil Heitz in Hamburg en route. During this session Dame Helen Gwynne-Vaughan (born Fraser) »gave a brilliant exposition of her observations and theories (on brachymeiosis in ascomycetes), and despite sometimes ra-

34 Westergaard and his co-worker O. Frydenberg in discussion (1959).

ther heated objections from a number of distinguished geneticists, there was no doubt for the genetically uninitiated, that she carried the day«. From thereon Westergaard was fascinated by the unique possibilities *Neottiella* offers for the study of chromosomes during meiosis in an ascomycete. Together with Jane Rossen (1966) he analyzed the cytology of meiosis with the light microscope and Dame Helen's error was uncovered. She had mistaken 32 chromatids for 32 chromosomes in the primary ascus nucleus. In this paper Westergaard also published his Feulgen cytophotometric data which revealed that the last DNA replication before the first meiotic division in this fungus takes place prior to karyogamy of the two haploid nuclei. Crossing-over at the four strand stage cannot occur in this case during DNA replication, as the two non-sister chromatids which will participate are located in two separate nuclei. By chosing the right object, it was for Westergaard a simple matter to bid farewell to the cherished copy-choice theory of crossing-over and the equally cherished precocity hypothesis of meiosis.

The other root for his studies on the mechanism of crossing-over is found in Westergaard's paper on chemical mutagenesis from 1957: »The next major break-through in genetics may come when the cytological approach can again keep pace with genetical analysis; when the electron microscope takes its proper place in genetical analysis, when we can study the submicroscopic basis for processes like inter-allelic crossing-over, forward and back mutations, gene conversion, etc. This day may not be far off«.

That this day has long since arrived is in no small degree due to Westergaard's work on the ultrastructure and cytochemistry of meiosis and the synaptinemal complex as well as the recognition that this universal structure of great evolutionary stability is a vector for fine pairing of homologous chromosomes and four-strand crossing-over. It comprised many, many happy hours for me to cooperate with Mogens in this venture for the last 13 years. It all started with the finding that the synaptinemal complex of the orange cup fungus, *Neottiella,* has a most pretty and regular structure. At the same time, the chromatin of the chromosomes is recognizable in the electron microscope at all stages of the mitotic and meiotic cycle. *Neottiella* has the advantage that one has to fix and treat the material in the forest or on the beach preferably on beautiful but out of the way spots in Denmark or on H. R. H. Prince Philip's polo grounds at Windsor Castle, where a special low chromosomic form occurs. This inter alia provides the necessary peaceful surroundings in which to discuss scientific problems and to sort them out. The ammoniacal silver ion reaction was tried, to learn something about the composition of the complex, but led to a very different discovery, namely that the lateral component of the synaptinemal complex is laid down at early meiotic prophase in the split between the two sister chromatids. This provides the solution to a problem debated for years at the light microscopic level, whether the leptotene chromosome consists of two threads or one thread: The leptotene chromosome is made up of two chromatids, but it appears frequently as a single thread because the two chromatids are connected by the lateral component of the synaptinemal complex. The other discovery that gave Mogens special satisfaction was the observation that a chiasma at early diplotene originates from a remaining piece of synaptinemal complex. We agreed immediately that this makes it almost unavoidable that crossing-over takes place inside the synaptinemal complex.

Since then Denise Zickler, Benjamin Lu, David McLaughlin, Bob Nilan, Chris Gillies, Larry Olson, Søren Rasmussen, Preben Holm, Herb Stern and Yasuo Hotta have cooperated for shorter or longer periods on aspects of meiosis to move us closer to Mogens Westergaard's goal that the cytological approach can keep pace with genetical and biochemical analysis. We are feeling lonely in continuation of this venture without his criticisms and his unconventional ideas which he pursued unperturbed in spite of our sceptical remarks, lonely without being trapped by one of his practical jokes and without being distracted by his newest interest in a piece of art, a book or some humorous story from the history of science.

As hard as I may try, it is not possible in this obituary to mention all the scientists who were personal and close friends of Mogens Westergaard. This is unfortunate since the contact with these friends meant so much to him and was paramountly important for his achievements.

I like to remember him in two ways: (1) There is a tradition at Copenhagen University that the examination requirements for undergraduate students are measured out in certain pages of one or several books. The examiner, of course, is not to ask anything which cannot be found on these pages. Confronted with this somewhat unacademic procedure, Mogens Westergaard decreed: This term we will require for the examination only the left pages of my textbook on genetics. (2) He walks over a spring meadow in full bloom, discovers a beautiful hybrid between *Orchis purpurea* and *Orchis militaris.* He kneels down to study and to photograph it.

35 Orchids in a meadow, photographed by M. Westergaard.

Martin Ottesen

by *Bent Foltmann*

Martin Ottesen was born 1920 in Ålborg, one of the larger towns in Northern Jutland. He went to the grammar school of the town (Ålborg Katedralskole) and in 1939 passed his »studentereksamen« (corresponding to the British A-level).

Like his predecessor at the Carlsberg Laboratory, K. U. Linderstrøm-Lang, Ottesen studied chemistry at the Technical University of Copenhagen and graduated with very high marks in 1944. One of his main topics was photography, a subject which ever since has remained one of his major interests.

The following two years he was employed at A/S GEA, Pharmaceutical Manufacturing Co., Copenhagen. During this period he published two minor papers on methodological aspects of pharmacology, but he became more and more interested in fundamental problems of biochemistry, so in 1946 he joined the staff of Linderstrøm-Lang at the Carlsberg Laboratory. Shortly afterwards, in 1947, he was married to Gerda Vinding.

Once he told me that originally his intentions were to study for a couple of years with Linderstrøm-Lang and then return to industrial research. But, like many others, he was trapped by fascinating problems and by the inspiring atmosphere of the Carlsberg Laboratory, to which he became a willing captive.

Ottesen's first investigations at the Carlsberg Laboratory concerned the binding of water to proteins in solution. In these days ovalbumin was used as a model substance in many experiments and one of the traditions of Carlsberg was that no resources ought to be wasted; thus after the physical-chemical measurements had been carried out, the protein samples were pooled and recrystallized. Now it happened that one day plate-like crystals were formed instead of the normal needle-shaped crystals of ovalbumin. This accident developed into one of the classical examples of serendipity, where a scientific breakthrough is initiated when the unexpected observation is made by prepared and interrogative minds. The sample of the plate-like crystals was not discarded as a failure. The questions were raised, what had happened and what made it happen?

Ottesen answered these questions with his characteristic experimental skill and diligent search; the answer deeply influenced his own career as well as our general understanding of the concept of »limited proteolysis«.

The plate-shaped crystals turned out to be a partial degradation product of ovalbumin and from the shape of the crystals it was called plakalbumin. The investigation of this process represents one of the first attempts to characterize a limited proteolysis, i.e. a proteolysis which involves the cleavage of only one or very few peptide bonds. Such small changes may profoundly alter the properties of the protein, as for instance when the zymogens of the proteases from the alimentary system are converted into active enzymes.

The initial degradation of ovalbumin consists in cleaving off a segment of 7 amino acid residues. The analyses of this peptide were published in four papers from 1951 to 1952, and this study is among the first to apply P. Edman's degradation method for sequencing an unknown structure.

To answer the next part of the question: what causes the transformation of ovalbumin into plakalbumin, Ottesen studied the conditions under which the pooled albumin samples were kept before crystallization. It turned out that the jars had become infected with a strain of Bacillus subtilis, and from this the guilty enzyme was purified; it was a new proteolytic enzyme, nowadays known as subtilisin. This investigation was car-

36 Martin Ottesen (1976).

ried out in collaboration with A. V. Güntelberg and published in 1954. About the same time Ottesen together with Charlotte Schellman showed that the enzyme was inhibited by di-isopropyl fluorophosphate so that it is now classified as a serine proteinase.

To pursue his research on the process of limited proteolysis from a quantitative point of view Ottesen needed better instruments than were immediately available in the laboratory. It may be difficult for the younger biochemists in the mid 1970's to imagine the experimental situation 20 years ago. Grants were small and the manufacturers were not yet competing to market as many sophisticated gadgets as possible. If a biochemist wanted to carry out a specific type of investigation he had to design an instrument which served his need. Together with the chief steward of the department Ottesen constructed cheap fraction collectors. The turn-table was moved by a weight suspended on a brass chain, the control unit was an electromagnet and a clockwork otherwise used to switch light off and on in shops windows – simple but efficient. In a fruitful collaboration with the instrument maker Ole Dich and later on with the Danish firm »Radiometer A/S« Ottesen also greatly improved the pH-stat of C. F. Jacobsen and J. Léonis, so that it became a sturdy and versatile instrument. Different types of pH-stats are now sold by several firms, but this original commercial type is still in use in many places all over the world.

The exact determination of base uptake during subtilisin degradation of ovalbumin enabled Ottesen to map out the consecutive steps of limited proteolysis leading to the formation of plakalbumin. In the initial reaction a particularly exposed peptide bond is hydrolysed creating a new C-terminal alanine residue (the corresponding new N-terminus was not identified, cf. Fig. 37). Even at this stage – after cleavage of one peptide bond – the protein changes its properties as far as solubility and crystal morphology are concerned. The reaction may now proceed via two routes. Either plakalbumin is formed by a one step reaction in which seven amino acid residues are cleaved off as a heptapeptide or by a two step reaction removing the same seven residues, first a dipeptide and then a pentapeptide. The resulting plakalbumin can be distinguished from ovalbumin by changes in solubility and electrophoretic mobility, but it retains its character of a globular protein.

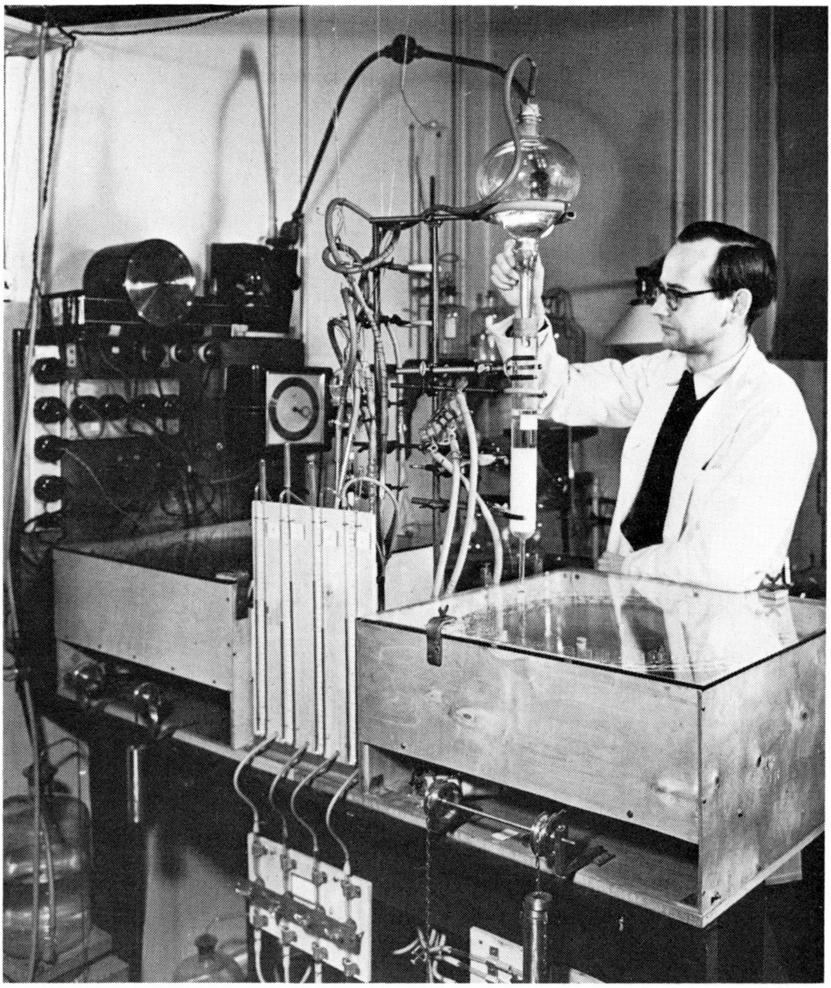

37 Ottesen at his home-built fraction collector for the analysis of peptides (1951).

In 1958 Ottesen summarized his fundamental investigations on the formation of plakalbumin as a case of limited proteolysis in a review for which he was awarded the doctor's degree at the University of Copenhagen.

Before this, the enzyme subtilisin had been used to investigate proteolyses of other proteins. Together with S. M. Kalman, K. U. Linderstrøm-Lang and F. M. Richards, Ottesen in 1955 showed that pancreatic ribonuclease could be inactivated by the action of subtilisin, but that active intermediary products presumably were present. Further studies of this reaction have extended our knowledge of the importance of the specific folding of the peptide chains in the formation of the active center in enzymes and of how the folding of peptide chains depends on the primary structure. To-day the action of subtilisin on ribonuclease is quoted in several text books of biochemistry as a typical example of limited proteolysis.

In parallel with his work on plakalbumin and subtilisin, Ottesen still maintained an interest in pharmacological problems and together with Margaret Gregory and E. S. Holdsworth he investigated the properties of a clinically active cyanocobalamin-protein complex.

In 1959 Ottesen succeeded Linderstrøm-Lang as head of the chemical department of the Carlsberg Laboratory. During his investigations of plakalbumin he never had lost interest in the physical-chemical properties of proteins, and from 1959 he conducted several projects concerning such aspects of protein structure. From 1960 to 1965 Ottesen published together with Aase Hvidt and several visiting biochemists (L. Kanarek, J. H. R. Kägi, A. Stracher and Giov. Di Sabato) a long series of papers on hydrogen/deuterium exchange between protein and solvent. Originally Linderstrøm-Lang introduced this method using density determinations as a measure for the exchange, and obtained valuable information about the availability of hydrogen atoms for exchange, i.e. information about how large numbers of the hydrogen atoms were shielded in secondary or tertiary structures in such a way that they could not be exchanged.

Gradually, however, the change in infrared spectra caused by the hydrogen exchange process became the method of choice for evaluation of hydrogen exchange in proteins. The CO-NH-bond has a characteristic absorbancy around 1550 cm^{-1} which disappears when hydrogen is exchanged with deuterium, thus providing an accurate tool for measuring the ex-

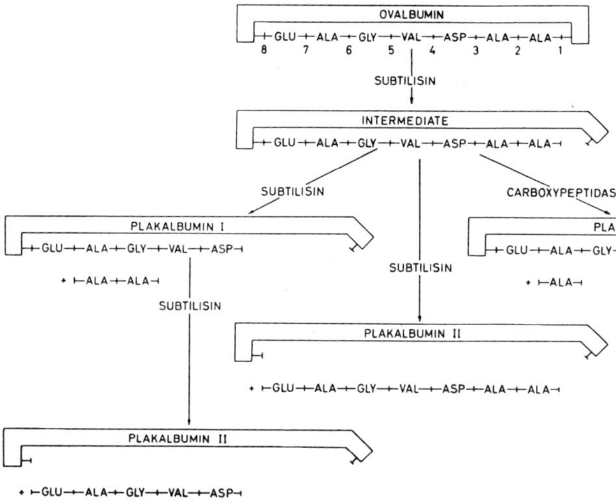

38 The first thoroughly studied example of limited proteolysis: the transformation of ovalbumin into plakalbumin (1958).

change. Later on tritium partly displaced deuterium for measurement of hydrogen exchange. This line of research on protein structure had occupied Linderstrøm-Lang during the last years of his life, and it has been ably and consistently pursued by Ottesen through many years. The entire development in the field was described in a comprehensive review »Methods for Hydrogen Isotope Exchange in Globular Proteins« which was published in »Methods of Biochemical Analysis« (1971).

In the course of his researches Ottesen has covered a wide field of problems in protein chemistry, but it seems that the proteinases from Bacillus subtilis have been his special favourites ever since their discovery.

By comparing the enzymatic properties of the enzyme produced by the original Carlsberg strain and a new strain of Bacillus subtilis used by the NOVO Industri A/S, Ottesen and A. Spector were able to show that the proteases produced by the two strains were separate entities. The next series of experiments on the subtilisin proteinases included a protease from a third strain used by NAGASE & Co., Ltd., Japan. These experiments comprised the analysis of the peptide pattern obtained after tryptic digestions of the three enzymes as well as determination of their amino acid compositions. From the results it was concluded that the original Carlsberg subtilisin was different from the other two which were closely related if not identical. This conclusion was confirmed when E. L. Smith and coworkers completed the determination of the primary structures of the subtilisins.

Together with the naturally occuring variants of subtilisin proteinases Ottesen and collaborators have also assayed the properties of artificial derivatives obtained by succinylation, carbamylation, nitration or iodination. These experiments were originally initiated to investigate the binding forces that confer a remarkable stability to the subtilisin molecules, but most interesting changes in the catalytic properties of the enzymes were observed after chemical modification. The exact interpretation of these results is not yet concluded, but the observations will certainly contribute to our understanding of the binding between the proteinase and its substrate.

In addition to these main topics of Ottesen's work examples of some sidelines ought to be mentioned. Together with H. Holter and R. Weber he has participated in the development of density gradient centrifugation as a means of separating cytoplasmic particles. Together with C. C. Bigelow and E. Frieden he has explored polyphenoloxidase and its effect on tyrosine containing proteins. Glutamine cyclotransferase was investigated together with M. Messer and an acidic protease from Mucor miehei together with W. S. Rickert.

This short survey of some of Ottesen's scientific achievements is based on papers of which he is an author. It would go beyond all bounds to describe all the scientific projects to which he has contributed directly or indirectly without appearing as an author of the final publication. The numerous acknowledgements in papers issued from the Carlsberg Labo-

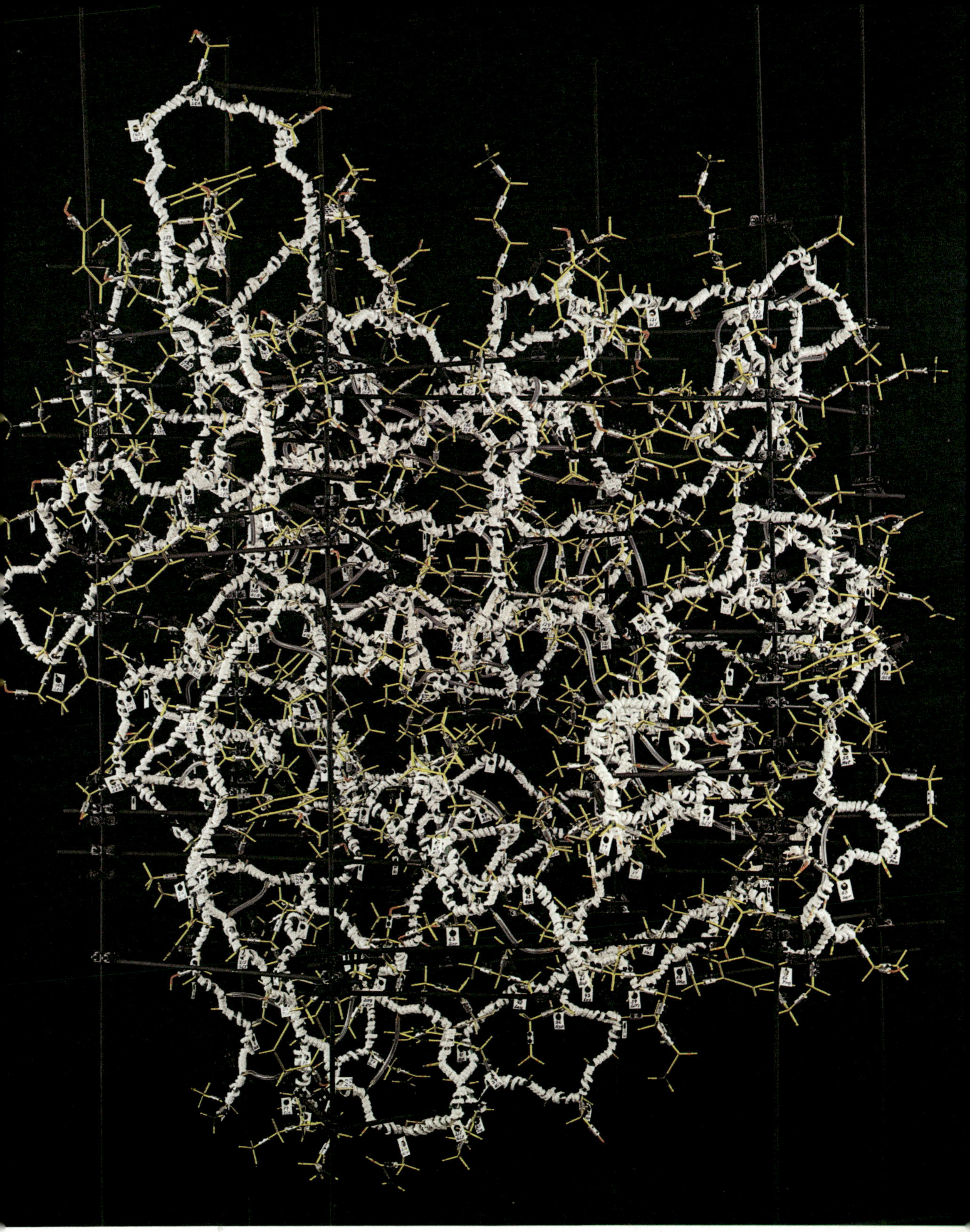

39 Subtilisin. Model showing the three-dimensional structure of a molecule of the crystalline enzyme Subtilisin Type Novo. The molecule consists of a simple peptide chain incorporating 275 amino acid residues. The model has been constructed by M. Ottesen in 1969, on the basis of X-ray investigations carried out by J. Kraut, University of California.

ratory over roughly 30 years speak for themselves. Personally I had the opportunity to work with Ottesen at the Carlsberg Laboratory from 1959 to 1963. It was a great experience to carry out the investigations on prorennin and rennin in such stimulating and inspiring surroundings.

In all problems which Ottesen has tackled he has worked with a penetrating insight and he has spared no effort to obtain reproducible and quantitative results. All reports or papers that have passed his desk have been carefully scrutinized to maintain the highest possible scientific standards. It must be emphasized, however, that Ottesen's advice has not been limited to guests and staff of the Carlsberg Laboratory. With open arms he has received scientists from other laboratories, listened to their questions and done his best to help them in solving their problems; members of clinical laboratories are especially indebted to Ottesen for many profitable discussions. The University of Copenhagen has also benefited greatly from his knowledge; since 1963 he has lectured in advanced courses on protein chemistry and served as external examiner as well.

Ottesen's interests have never been limited to narrow biochemical problems. From his youth he has been engaged in problems of human society with a deep sense of responsibility. In the late 1940's he was among those who took the initiative to establish a union of Danish scientists. In the 1950's he participated actively in the anti-atomic-warfare campaign and since 1971 he has been a member of the board of the Danish association against drug abuse.

Ottesen has held several professional honorary offices and many honours have been bestowed on him. He has been president of the Danish Biochemical Society and he has been a member of the Assembly of the International Union of Biochemistry. He has also been a member of the editorial boards of Acta Chemica Scandinavica and of Biochimica et Biophysica Acta. In 1959 Ottesen was awarded the Niels Bjerrum Memorial Medal, in 1960 he was elected a member of the Royal

Danish Academy of Sciences and Letters, and in 1962 of the Academy of Technical Science. Since 1966 he has been chairman of the board of the Linderstrøm-Lang Memorial Medal Foundation.

Ottesen has travelled much, but he has never been abroad for an extended period of time. It simply was not necessary. He came to the Carlsberg Laboratory at an age when many young biochemists would ordinarily go abroad, but he had no need for such explorations, since there he was able to meet the people that others must seek out by travelling. Most of the outstanding protein chemists of our days have visited the Carlsberg Laboratory and many have worked there for one or more years. Ottesen has learned from them all and paid back with his personal knowledge and great experimental skill. As the first-assistant of Linderstrøm-Lang and later on as head of the chemical department he carried a heavy burden in helping many a migrating bird to find a temporary nest at the Carlsberg Laboratory.

In one of the introductory paragraphs I mentioned that Ottesen originally had intended to return to applied research. However, he has never needed to abandon such research altogether, since the professors at the Carlsberg Laboratory have traditionally engaged in such problems of applied research as were necessary for the breweries. Like his predecessors, S. P. L. Sørensen and Linderstrøm-Lang, Ottesen has been a member of the scientific advisory board of the Carlsberg Breweries since he became head of the chemical department in 1959.

In another chapter of this book the story is told how the Carlsberg Laboratory has come back to the breweries. This means that in a close way Ottesen is able to unite his two interests, and we can only congratulate him for the present change in the material conditions of the Laboratory. During its centenary year the chemical department is expanding to larger and renewed laboratories while the research goes on, flourishing as ever.

V

Physiological department. Biographies.

Rasmus Pedersen

1840–1905

by *D. Müller*

Rasmus Pedersen was the first Danish plant physiologist, and the first director of the physiological department of the Carlsberg Laboratory.

Rasmus Pedersen was born on May 29th 1840 in Marslev, between Nyborg and Odense, as the son of the farmer Peder Chr. Mortensen (1802–54) and Ane Rasmusdatter (1806–96). He worked on his father's farm until his fifteenth year, but when his father died a small legacy enabled him to study. He matriculated from Odense Cathedral School in 1860 and began to study medicine, as well as devoting some time to mathematics and botany. He was particularly impressed by the teaching of Professor P. L. Panum (1820–85), the first professor of human physiology in Denmark, who, up to the time of his death, remained his mentor and faithful supporter. It was Panum who caused Rasmus Pedersen to study plant physiology. In 1871 he was awarded the gold medal of the University for an essay on »The scientific and practical applications of electromagnetism«, and the Royal Danish Academy of Sciences and Letters awarded him a consolation prize for two essays (in Danish), »The part played by cleavage of the growing point

40 Rasmus Pedersen (1904).

in the branching of flowering plants« and »The development of the inflorescence in the spurge (Euphorbia)«.

His diligent studies were rewarded by funds for studying in Germany from 1872 to 1876, first in Leipzig with the plant anatomist and phytopaleontologist August Schenk (1815–91), and later in Würzburg with the founder of recent developments in plant physiology, the brilliant and indefatigable Julius Sachs (1832–97). When Rasmus Pedersen returned he was the first scientifically trained plant physiologist in Denmark, and had also an extensive training in plant anatomy and human physiology. He was therefore the obvious choice for the post of head of the physiological department of the Carlsberg Laboratory, which had just been founded by the brewer J. C. Jacobsen (1811–87). There is no doubt that Panum (who was a member of the board of trustees of the laboratory), J. C. Jacobsen, and Rasmus Pedersen himself all anticipated that as head of the physiological department, which was a purely scientific post without teaching responsibilities, he would be able to live up to the problems of the laboratory.

In this post he completed the following three investigations, which were published in vol. 1 of Meddelelser fra Carlsberg Laboratoriet (now Comptes Rendus des Travaux du Laboratoire Carlsberg) (1882):

(1) Investigation of the factors influencing the reproduction of the bottom yeast form of Saccharomyces cerevisiae.

(2) Investigation of the effect of introducing atmospheric air in fermenting wort during fermentation.

(3) Investigation of the effect of temperature on the production of carbon dioxide in barley seedlings in the dark.

These three publications are both critical and relevant to the aims of the Carlsberg Laboratory; the results were correct and were based in part on new methods. However, J. C. Jacobsen was not satisfied with Rasmus Pedersen's working methods, and he was therefore dismissed at the end of 1877.

This dismissal can be deemed fortunate for two reasons. It was fortunate because Rasmus Pedersen was not an experimental physiologist, and could hardly have succeeded in organising fruitful experimental research. It was also fortunate because it compelled him to return to his original aim, the institution of teaching and research in plant physiology at university level, as a scientific basis for plant cultivation; he had this in view during his medical studies and in the four years which he spent in Germany. Perhaps it was his farming origins which led to his wise and for-sighted emphasis on the importance of plant physiology as a basis for cultivation.

With the support of Professor Panum and others a university lectureship in plant physiology was created in 1879 and converted to a professorship in 1887; this post was held by Rasmus Pedersen for 26 years, until his death in 1905. This was the first chair of plant physiology in Denmark, and the second in the world, being preceded only by the chair in Vienna. The other European plant physiologists, such as J. Sachs and W. Pfeffer, were professors of botany. The establishment of this first chair of plant physiology was Rasmus Pedersen's greatest achievement in the service of the subject, and for this alone he will always be remembered in the history of Danish botany.

However, unfortunately no research work in plant physiology emerged, apart from thorough studies of the literature. In a letter dated January 13th 1881 to the plant pathologist Emil Rostrup (1831–1907), Eugenius Warming (1841–1924) writes »Rasmus Pedersen gives good lectures in physiology, but does no experimental work«. In the period 1890–97 the first and only assistant in the laboratory of plant physiology was Hjalmar Jensen (1865–1948), who had to organise single-handed the practical classes in plant physiology, while his evening task was to produce (with one of the primitive gelatine hectographs then in use) ten copies of the copious tables which Rasmus Pedersen used to supplement his lectures; there were no slides or lecture demonstrations.

Rasmus Pedersen was crammed full with erudition, and his lectures were based on a far too comprehensive range of reading. His favourite, and only pupil, J. Chr. Bay (1871–1962), who also had worked at the Carlsberg Laboratory (with E. Chr. Hansen), and who emigrated in 1892, ending up as librarian at the John Crerar Library in Chicago, took with him to the U.S. a fair copy of notes from Rasmus Pedersen's lectures amounting to 4000 pages.

It was probably Rasmus Pedersen's intention to write a history of plant physiology, but of this great project only the first part appeared, »Plant nutrients. A historical introduction« (pp. 343, Copenhagen 1883). This is a first-rate book, worth reading to this very day, since the author has had the excellent idea of using the actual words of earlier workers in the form of long and well-chosen quotations, thus ensuring that the book will never be out of date.

Rasmus Pedersen framed the regulations for the Master's Degree in plant physiology, but the only candidate he himself succeeded in graduating (in 1897) was Frederik Weis (1871–1933), later assistant at the Carlsberg Laboratory.

Rasmus Pedersen was a gifted person with a receptive, historical and critical mind: he was weighed down by erudition and had a gloomy temperament. The bacteriologist Carl Jul. Salomonsen (1847–1924), a friend of his youth, wrote in his obituary; »He produced little, and in particular he was ill suited for experiental work«.

Strangely enough we know nothing about Rasmus Pedersen's attitude to the greatest controversy of that time, Ch. Darwin's theory of evolution. A few words in Bay's biography (1919) could be taken to mean that he was unsympathetic to the theory of evolution, as was the leading Danish zoologist Japetus Steenstrup (1813–97). From 1890 until his death Pedersen lived in one of the two large professors' residences in the Botanical Laboratory at Gothersgade 140. The other residence was occupied by his contemporary Eugenius Warming (1841–1924), who was a fanatical Lamarckian.

Rasmus Pedersen was tall and heavily built, but adversity

and defeat had rendered him snappish and troublesome Combined with long nights of poring over books this undermined his health at an early stage. He died on May 2nd 1905 leaving a widow Frederikke (née Müller) whom he had married in 1888 and who died in 1927 at the age of 79.

Emil Christian Hansen

1842–1909

by *Alb. Klöcker**

Emil Christian Hansen was born on May 8, 1842 in Ribe in Southern Jutland (Slesvig), a son of Joseph Christian Hansen, a house-painter, and his wife Ane Cathrina, née Dyhre.

The elder Hansen had led a most adventurous life. From his subtle hints he was very likely an illegitemate child of distinguished heritage; yet he did not speak about it when his children grew older. He was born in Copenhagen, brought up as a foster child by an old woman and later adopted by the headmaster of Jonstrup Seminarium. Shortly after his confirmation he decided to become a sailor but after only 7 months he returned to become a painter's apprentice in Odense. After 3 years he went abroad, painting boats in Bremerhaven and theatre decorations in Karlsruhe, where he even appeared on the stage as a member of the chorus. He travelled mainly in Southern Germany, where he worked as a landscape- and portrait-painter as well as house-painter.

During the early 1830's Joseph volunteered to join the French Foreign Legion and participated in the war against Abd-el-Kader. In 1835 the Legion went to Spain to fight against Don Carlos, and Joseph was wounded in 1837. In France he later took part in a number of insurrections and was subsequently deported. During the winter of 1838 he arrived back home in Slesvig, barefoot in the snow. After a short stay in Copenhagen he returned to Haderslev to work as a house-

41 E. Chr. Hansen in the old Carlsberg Laboratory. (Painted 1896–97).

* *Written in 1910.*

painter. He was generally known as »the crazy fellow from Copenhagen«. He became engaged to Ane Cathrina Dyhre whose father had been a tailor, an innkeeper and a farmer. Joseph Hansen then went to Ribe where he was admitted to the Painters' Guild, and where the marriage took place on June 19, 1841. Here their son Emil Christian Hansen was born.

The boy received his first education in some small private schools for children. From the time Emil was able to walk, his father took him along on strenuous walks in the countryside, no matter what kind of weather, winter or summer. They often returned late at night after the father had consumed a number of drinks of coffee spiced with aquavit at the local inns. He usually had to carry the tired boy, and it often happened that they lost their way in the dark. The father would tell him stories from the wars and his other adventures, inevitably concluding with the words: »a person can do everything as long as he has the will to do it«. The boy heard this so often that it left a deep impression on him. Once, when Emil was 5 years old, they were walking towards home late one night, pulling a cart full of paint pots. In the middle of the road the cart broke down. The father took as many pots as he could carry and left the boy alone in the pitch-dark night to guard the rest till the father returned. At another time the father ordered Emil to jump into a deep pond. The boy was too scared not to obey but since he could not swim, he sank to the bottom, and the father had to jump in to save him.

The years of the war, 1848–50 (the Holsteiners' revolt and Prussia's intervention) impressed the boy very much. Since there was very little house-painting to do, the father worked as a spy for the Danish army.

After the war, Emil started in a municipal school. He did not learn very much but he did pick up various tricks such as how to play truant. Later he entered a »paying school«, i.e. private school. At the beginning he had difficulties in keeping up with the other pupils, especially in arithmetic but soon he was at the top of his class. Feeling alone and shy, and already knowing the seriousness of life, his thirst for learning was enormous. His parents became poorer and poorer, and there were still more children. They could not even afford a hot meal every day. During summer-time all went rather well but the winters were hard. The father and Emil would then collect scrap iron and lead from the dumps to sell. They also earned a little by decorating roller blinds which they sold in Slesvig.

As Emil grew older, he would earn some money by copying for some attorneys or by working as a delivery boy, and during his last years in school by tutoring the duller pupils.

Whatever time was left he would spend reading, and he was especially interested in natural history. He happened to get hold of a French and a Spanish grammar book, and so began studying these two languages.

Emil had certainly inherited some of his father's gypsy habits, and it was reflected by his alternating desires towards education with most different goals. When there were visiting theatre companies, the boys in town got free tickets in return for carrying the costumes from the lodgings to the theatre. This led Emil to the idea of becoming an actor. He rehearsed a number of parts, succeeding in performing as one of the best actors on a local community stage. In spite of this, his studies were not diminished, and it was still his greatest hope to enter grammar school. The family's financial situation did not, however, make this feasible.

During his last years in school, Emil Hansen worked as a transcriber to the headmaster of the grammar school and also helped with book-keeping. The headmaster wanted him to enter his own school but the father was against it and wanted Emil to go into commerce. Half a year before his confirmation, still not quite 14 years old, he became apprentice in a grocery shop. The apprenticeship was supposed to run for 5 years, during which time he would receive room and board, and footwear; clothing his parents had to supply. As was to be expected in those days, he was mistreated from the beginning by the shop assistant and the older apprentice. The relations improved somewhat, after Emil one day gave the older apprentice a real thrashing. As he soon realized that he was no good at commerce, he tried to get out of it. The parents resisted very much as it was considered most reproachable to desert an apprenticeship. Emil now got the idea of becoming a sailor and wrote to his uncle, a shipowner who quickly made him change his mind. At that time the post in the grocery shop had become so hopeless that he one day went down to the sea and swam out very far, only wanting to drown. A great fear overtook him, and with his last strength he managed to reach the beach. Finally, after another 6 months at the grocery, his parents agreed upon his leaving it. Emil now decided to become a pharmacist or to work in an office where he could study law in his spare time. However, nobody wanted to have anything to do with him, and many people considered him a little crazy. In fact, he was held up as a warning example to young people.

The only solution thus was to take up his father's profession. So he eagerly started to paint houses, and was soon a journeyman painter. Nor did he just paint houses, he also copied paintings and did original drawings and scenery paintings as well.

On April 10, 1860 the police issued travel-papers to house-painter Emil Chr. Hansen. The day after he travelled by foot, with only 5 Rigsdaler in his pocket, to Haderslev. As no jobs were to be found, he continued to Aabenraa, where he got a job. When this was finished, he started fending his way southwards. Between Rendsburg and Neumünster he joined a gypsy family. They wanted him to stay with them and go to Cologne but he declined and wandered north again. Finally he reached Ribe, with torn clothes and worn-out boots.

It was winter-time, and there was no work. He spent his time copying the large painting »Christ's Walk to Golgatha« by professor Jørgen Roed, in St. Catherine's Church. One teacher in town liked the copy and wrote about it in the local paper. So Hansen sent his painting to Copenhagen; he himself walked to Kolding and from there went by boat to Copenhagen.

Now he wanted to try his luck as an artist. He looked up professor Roed and showed him the copy but it was not too well received. Hansen lacked schooling and the only advice he got was to start all over again and learn to draw. So much for his hopeful expectations. The painting was sent back to Ribe, where it was bought by the widow of a local parson.

Since Hansen had been unsuccessful in finding a job in Copenhagen, he now went to Roskilde. No work. Then to Holbæk where he had better luck. From there to Ringsted, and then back to Copenhagen with his meager savings. He immediately tried to be enrolled at the Royal Academy of Arts but was rejected. This finally induced him to give up the idea of becoming an artist. Instead he decided to be a school teacher. He earned enought to support himself during the necessary studies by drawing and tutoring. He also did some accounting for a minor junk-dealer. It was indeed difficult to earn enough, and he toyed with the thought of wandering through Germany to Northern Italy to volunteer to join Garibaldi's army. Suddenly, however, he received a notice that he was accepted as tutor for the children of the administrator of the castle Holsteinborg in southern Sealand. By sheer accident his application was chosen among a great many similar ones.

It was an incredible stroke of luck for him to get this position. He now eagerly began to study natural sciences in his free hours. He formed a lasting friendship with the agricultural counsellor P. Nielsen who at that time was a teacher in the village of Ørslev near Holsteinborg. Here Hansen received his first guidance in identifying plants, and soon he had a comprehensive herbarium and insect collection.

From now on his studies for the final examination were

carefully planned; first he was going to get his teacher's certificate and then he wanted to pass »Artium« (the final of senior high school). Count Holstein's house-tutor helped him with Latin. In 1864 he went to Copenhagen to take more courses at Blaagaard's Teachers Training College, and during the war of 1864 against Prussia he passed the examination and received his teacher's certificate. He then reported to the draft board but just as before, when he had reported to reserve officers' school, he was rejected, since he was nearsighted and wore glasses. He returned to Holsteinborg and continued his tutoring, at the same time preparing himself for the first part of the »Artium« degree which he passed in Copenhagen in 1865. Thereupon he took another tutoring job, this time on the island of Fynen, mainly studying botany in this period. For a while he considered becoming a veterinarian or if possible a physician. As he had worked excessively hard while studying for his examinations, his health had suffered, and unfortunately he had not enough money to pay for the continuation of his studies. Consequently he applied for a Monrad Scholarship grant, which would pay a teacher's further education for 3 years at the Copenhagen Technical University. He was lucky enough to get it, and he returned to Copenhagen in 1866 to resume his studies. He was still in poor health; he came down with typhoid fever and had to stay in bed for nearly 4 months. His fellow students were now far ahead of him, and so he had to decide whether to resume his studies in the same arduous way as before or give it all up. He chose to continue and this time his health improved. In 1869 he passed his final examinations with high marks.

To earn a little extra money in these hard years he wrote several stories, sometimes under the pseudonym »E.C.«.

He got a job as a teacher of natural history and English at Schneekloth's grammar school, while at the same time tutoring medical students and the students at the Veterinary and Agricultural University. The disciplines were zoology and botany.

Simultaneously he passed in 1871 the second part of »Artium«, and half a year later he took the degree in philosophy at the University of Copenhagen. Among his teachers the zoologist J. C. Schiødte had made a great impression on him, so he considered specializing in zoology. However, the following year he became still more attracted to plant physiology but before he became really involved in this, professor Japetus Steenstrup employed him as assistant in investigations of peat-bogs.

At one point during these studies Hansen's conclusions regarding one of the peat-bogs in Northern Sealand, »Femsølyng«, were opposite to those reached by Steenstrup. Among other things, Hansen found that the beech and ashtree belonged to the oldest layers. Steenstrup immediately went to »Femsølyng« himself to verify Hansen's claims – and suggested that Hansen publish his observations as a chapter of his own in a larger paper which Steenstrup planned to publish through the Royal Danish Academy of Sciences and Letters. Gratefully Hansen accepted but the work was never published, nor was any manuscript to be found after Steenstrup's death.

Hansen did, however, give a lecture on his observations in Dansk Naturhistorisk Forening (the »Danish Natural History Society«), the chairman of which was Steenstrup, and the lecture was published in its »Scientific Communications« in 1873.

Hansen continued his investigations of peat-bogs over a span of years but they were never published.

The above mentioned discovery in combination with the lecture resulted in a cooling off in the relationship between Steenstrup and Hansen, according to Hansen who even suspected a certain degree of antagonism. This supposition was scarcely justified, since it was partly thanks to Steenstrup's recommendation that Hansen later received his appointment at the Carlsberg Laboratory.

In 1874 the University of Copenhagen proposed as the subject of a prize essay: »Give a systematic description of Danish fungi, depending on mammalian dung which in no way has been interfered with by man«. Hansen's paper on this subject was awarded the Gold Medal of the University in 1876. This was, indeed, a great honour, and the judgement of the committee ends as follows: »We consider this monograph, based upon a tireless and circumspect work, to be the first exhaustive treatise on the subject proposed, and we find it fully worthy of the prize.« Part of this work was published in the Scientific Communications of Dansk Naturhistorisk Forening in 1876.

In 1877 Professor P. Panum suggested that he and Hansen jointly should do research on fermentation. This marked the beginning of Hansen's interest in yeast physiology. Simultaneously Hansen applied for permission to defend a doctoral thesis although he had not obtained a Master's degree. His application was granted, and he then started – in Panum's laboratory of physiology – the studies of yeast fermentation physiology which were to be fundamental for all later works. He began by examining the organisms occuring in beer and beer-wort.

At this time the director of the Ny Carlsberg Brewery, Carl Jacobsen was looking for a scientifically trained young man who would perform daily microscopic observations of the beer during an extended period of time. On the recommendation of Steenstrup the post was given to Hansen. However, at the end of 1877 the position as head of the Physiological Department at the Carlsberg Laboratory became vacant, and Hansen, aspiring to it, got permission from the old brewer J. C. Jacobsen to work in the department. Thanks to the excellent working conditions, including living quarters right next to the laboratory, Hansen started work on the project which he hoped to use for his thesis. He was able to finish it within a year, and in the beginning of 1878 he handed in his thesis: »About organisms in beer and beer-wort«. He received his degree April 25, 1879.

To complete his work for the doctor's degree Hansen had

42 The young E. Chr. Hansen (c.1875).

worked day and night, and J. C. Jacobsen was most interested in his progress. His frequent visits to the laboratory did, however, disturb Hansen to the extent that he finally asked Jacobsen to stay away and let him work in peace. Jacobsen rushed out the door, and Hansen was afraid he had gone too far and would be dismissed; Jacobsen, however, talked jokingly about the incident later on.

Starting January 1, 1879 Hansen was made acting head of the Physiological Department of the Carlsberg Laboratory with full salary, and as of October 1 the same year the appointment was made permanent.

On March 30, 1880 he was married in the church at Holsteinborg to Miss Mathilde Melchior who until then had been matron at Gamle Carlsberg Brewery.

1883 was probably the most important year in Hansen's life, the year when he announced his system of pure yeast cultures and when Gamle Carlsberg Brewery did the first experiments with his pure yeast strain.

Hansen had conceived the idea that the main reason for the frequent occurrence of aberrant brews resulting in unsaleable beer was not bacterial infection (as L. Pasteur had assumed) but contamination by »wild« yeasts. He therefore devised a method by which he could isolate a single yeast cell and propagate it into an absolutely pure culture. J. C. Jacobsen at first had his doubts about Hansen's ideas but had to give in when confronted with the results of the first experiments, and from then on he watched with the greatest interest Hansen's attempts to separate a pure and useable yeast strain from the mixture employed in brewery practice.

Finally Hansen succeeded in culturing enough of the pure yeast, »Carlsberg's bottom yeast no. 1«, to use it in the brewery's fermentation wats. The new yeast was used for the first time on a production scale on November 12, 1883 at Gamle Carlsberg. The result was a most convincing success. Jacobsen's last doubts disappeared, and he used his great authority to speedily communicate the reforms as widely as possible, among other things by sending notices to a number of famous breweries, offering them the pure yeast for experimentation.

Hansen's aim had also been to provide Gamle Carlsberg

43 Apparatus for propagation of pure yeast from single cells, as used in the Ny Carlsberg Brewery (1888). (Compare fig. 93).

with additional income through his inventions, and he therefore suggested that Jacobsen might establish a factory to produce pure strains to supply users all over the world with pure yeast. But Jacobsen did not agree; whatever was discovered at the Carlsberg Laboratory belonged to the entire world. So, Jacobsen again in fact gave away large sums of money.

In 1882 Hansen arranged his first course at the laboratory, and many famous scientists participated. These courses continued over some years, and numerous chemists and botanists received information about the discoveries made at the Carlsberg Laboratory.

In 1902 Hansen celebrated the 25th anniversary of his appointment at Carlsberg. A large sum had been collected through an international subscription, and this, together with a gold medal was presented by the brewery director, dr. Carl Jacobsen as representative of the Committee, with these words:

»25 years ago, you, Professor Hansen began your studies at the Carlsberg Breweries and at the Physiological Department at the Carlsberg Laboratory. Your merits are well-known and highly esteemed, not only in your own country but throughout the world. Your discoveries have resulted in enormous progress in the study of the biology of yeast organisms, and your teaching of how to use purer yeast species and strains, selected at will according to exact plans have resulted in extremely important reforms within the fermentation industry. We, the signers of this address, are together to-day to present you with a gold medal as an honour and appreciation of your ingenious and pioneering studies during these last 25 years.«

At the beginning of 1909 Hansen's wife was seriously ill, and soon he became ill too. He suffered much, and gradual loss of hearing depressed him very much. He hoped, however, that they could both recuperate in the countryside but as his wife's condition deteriorated, he decided to go alone. He had reserved a room in a pension at the seaside resort of Hornbæk but feeling too sick and weak he went to a sanitarium instead. Upon his arrival he already had a high fever and went to bed – never to get up again. He weakened rapidly and died on August 27, 1909.

Before his appointment at the Carlsberg Laboratory Hansen's work was mainly botanical with special emphasis on the morphology and biology of the higher fungi. An experimental physiological approach did not develop until early in 1878 when his working conditions improved considerably as a result of his appointment. From then on he threw his energy into the study of microorganisms at large (bacteria, yeasts, moulds). After a period of orientation, he placed the main emphasis on the Saccharomycetes. This introductory period spanned from 1878 to the end of 1879. After this, the second period starts and lasts from the end of 1879 to approximately 1889.

During the second period the target for his research became the design of techniques for the study of the physiology of fermentation, studies of Saccharomyces species and their variants as they occur in industry and in nature, as well as their life cycle in nature.

His investigations were at that time mainly based on the point of view that the Saccharomyces species studied were constant types. Yet, at the same time Hansen had a keen eye for the existence of variants. His most important predecessors in this field, M. Reess and L. Pasteur, had both started on quite different principles. In the Pasteur approach all yeast cells which caused alcoholic fermentation were simply referred to as Saccharomyces. No differentiation between Saccharomyces species and non-Saccharomyces species was attempted;hence species belonging to entirely different compartments in the system were pooled together. Under such circumstances any exact scientific study of the species was ruled out; a botanical treatment was not even considered by Pasteur, as his education and talents went in other directions.

Reess on the other hand had realized the occurence in nature of a group of yeast cells distinguished by endogenous spore formation. Thus, he classified Saccharomyces as a separate group of yeast but since he was not an experimentalist, he was content with the descriptive morphology and proposed, as was the custom of botanists at that time, species according to shape and size without realizing that these characters vary greatly with conditions of culture. Hansen's approach was based on the realization of the eminent importance of growth conditions for the physiology and morphology of yeasts. He considered his task the opening of new roads, not the description of new species. His work thus resulted in contributions to the general morphology, biology and physiology of the Saccharomycetes, and it has been said by one of his colleagues, K. Windisch, that Hansen's chief merits lay in the replacement of foggy theories and loose speculations by clear examples and by facts with practical consequences. It was precisely with practical consequences as an objective that Hansen, day after day, year after year embarked on the experimental study of Saccharomyces. His experiments of 1880 laid the foundation for his doctrine about pathogenic yeasts. In 1881 he had worked out a precise culture method for isolation of pure strains, which he later improved.

Pasteur's view on fermentation techniques was again different from Hansen's. Pasteur had discovered that some of the diseases of beer and wine were due to bacteria, and his objective became therefore the elimination of contaminating bacteria. For this purpose he used cultivation in sucrose solution containing a small amount of tartaric acid. However, Hansen showed in an extensive series of investigations that while this device eliminated the bacteria, it also enhanced the development of pathogenic yeasts. In practice, therefore, Pasteur's device was useless for the simple reason that – at that time – one could not distinguish good brewer's yeast from pathogenic yeast since these two categories were as yet unknown.

Hansen started from new premises and new methods with abolutely pure cultures from one single cell, using botanical as well as physiological criteria, and thus arrived at valid definitions of pathogenic yeasts and cultured yeasts. The practical consequence was the attainment of pure cultures of pre-selected species with known fermentation properties.

When Hansen in 1883 presented his new theories to the fermentation industry the opposition evoked was to some extent justified. It was only after several years that he was able to give complete proof for the correctness of his postulates. Thus it did not suffice that Hansen started from the right ideas;

there remained a struggle which had to be conducted for years, both with theoreticians and practitioners, before his reforms would become part of practical life.

In 1889 begins the third period of Hansen's work. As previously mentioned, Hansen's studies on Saccharomyces were largely governed by the assumption that this yeast preserved constant characters. From now on the experiments were directed towards the variations of these characters.

The first significant variants were developed by Hansen in 1889. Under the influence of a specific method of cultivation, Saccharomyces loses the ability to form spores and is unable to regain the lost faculty, regardless of seemingly favorable culture conditions over several generations. Hansen's purpose with these complicated experiments was twofold: partly to explore the laws for variation, and use them to illuminate Ch. Darwin's hypothesis, partly to achieve a transformation of yeast for industrial purposes. The practical result achieved consisted mainly in the production of variants with more vigorous growth or with changed fermentation characteristics. These investigations are thus useful for the manufacture of compressed yeast and spirits.

In his later years Hansen mainly continued his studies of variations and tried to clarify some basic problems: Is the appearance of the sporeless variants under certain environmental conditions caused by true cellular transformation, or is it merely a result of selection of certain favoured cells? He arrived at the conclusion that transformation occurs, but gradually, resulting in an unchangeable variant.

Of equal interest are Hansen's experiments with top and bottom yeasts. Over the years it had been disputed heatedly whether the two forms were truly independent or whether transitions from one to the other did occur. As a result of Hansen's experiments it became clear that among the cells of a typical bottom yeast one would encounter some which gave true top fermentation and vice versa. The magnitude of Hansen's effort is illustrated by the fact that in one case it was necessary to isolate 9948 top yeast cells in order to encounter a single bottom yeast cell. The main result of these investigations was that the two physiological forms, top and bottom yeast, are not independent but that one can develop from the other. Hansen refers the variants to the category of »mutations«.

In the following years a significant part of Hansen's investigations were dealing with the conditions governing budding and sporulation. He demonstrated that the cells are able to form spores directly, without prior budding. The spore itself is able to give rise to a spore parental cell, thus by-passing intermediate stages.

As mentioned previously, during the earlier years of his studies on fermentation organisms, Hansen made attempts to follow the life cycle of these organisms in nature. He was successful in clarifying this with regard to Saccharomyces apiculatus, a species readily recognizable on account of its lemon shaped cells. The natural habitat for this yeast turned out to be sweet juicy fruits. Such fruits fall on the ground or are eaten by birds or insects, and through their droppings end up in the soil; there the yeast stays alive through winter till the next fruiting season. Hence this species is encountered in particular in the soil under fruit shrubs and fruit trees.

The other genuine yeasts by and large have similar cycles. Yet, the last of Hansen's investigations in this field revealed that the Saccharomycetes also have other growth localities although not as favorable as in the sweet juicy fruits. He found the yeasts rather constantly in soil far from the so-called primary foci, fruit gardens and wine slopes. These other locations he called secondary foci. They are of a very different nature: grass fields, lichens and mosses on tree trunks, etc. The farther away from the primary focus, the more sparse the occurrence of yeasts; higher up in the mountains, for instance, they disappear completely. When death interrupted Hansen's work, he

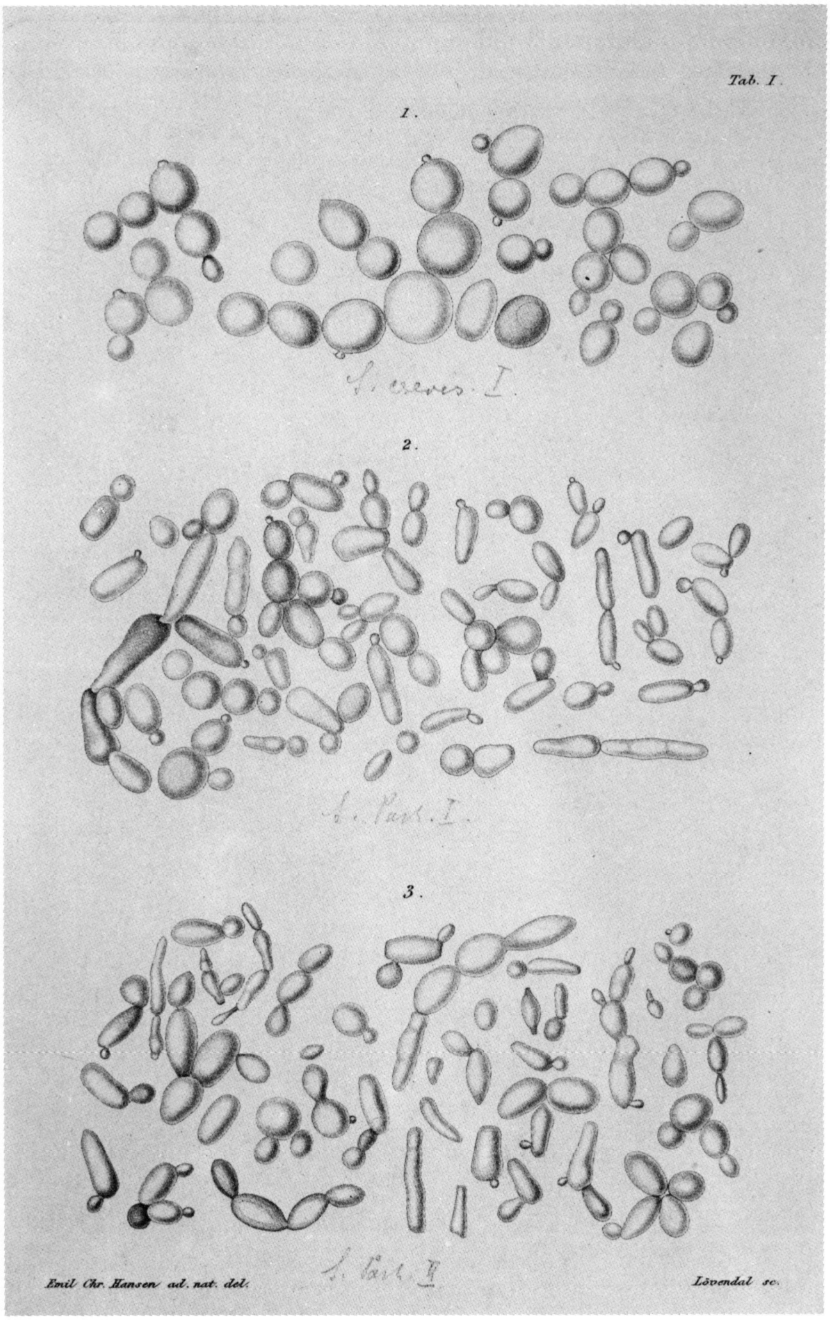

44 Yeast cells drawn by E. Chr. Hansen
1. Saccharomyces cerevisiae I; 2. Saccharomyces Pasteurianus I; 3. Saccharomyces Pasteurianus II (in Hansen's own hand).

was extending his investigations in this field to arctic and tropical regions.

Finally it should be mentioned that Hansen during his last years established a framework for the systematics of the Saccharomycetes to provide some order in the chaos which the descriptions of the many species gradually had produced. This framework has since been generally adhered to.

During his last years Hansen also reverted to the study of the higher fungi. It was especially the biology of sclerotium-forming Coprinus species (»Shaggymane« mushrooms) which he tackled in a continuation of the work of his youth (for which he, as mentioned, had received the Gold Medal of the University). In these investigations he elucidated the development of 3 characteristic species and thereby gave typical examples of the development of Coprinus in general. It was demonstrated that light promotes motility in these mushrooms and regulates the ejection of spores. With respect to Coprinus stercorarius, he made the important discovery that the spores are enclosed in an envelope with the features of a sporangium. Concerning the higher mushroom flora in the Copenhagen area, Hansen has left a number of notes.

Apart from science, Hansen had other interests, especially art and the history of art. His frequent trips abroad often took him to places where great art treasures could be found, and there he eagerly and diligently cultivated his love of the arts, just as he was an eager and understanding connoisseur of the art treasures of Denmark. He had collected a large library, mainly in science but a considerable part in art history. He also left a sizable collection of etchings, prints and drawings. Over the years he collected a number of framed portraits of famous scientists, which the Carlsberg Laboratory received after his death together with a large collection of reprints and books on these subjects. He himself had collected a great deal of information for a biography of Otto Friedrich Müller, whom he

45 Hansen at his microscope in the new Carlsberg Laboratory (c. 1900). Foreground: the »Hansen box« for the sterile handling of yeast. Background: Hansen's office with desk.

considered Denmark's most outstanding natural scientist. In his youth he wrote about Niels Stensen and Jan Swammerdam in »Historisk Archiv«.

Above all, Hansen was a scientist. He considered even the most ordinary events from a serious scientific point of view, and this attitude, combined with his peculiar ponderosity, which marked both his mind and his appearance, resulted for instance in his being unable to appreciate a joke: he took everything seriously. He seemed to exude this scientific seriousness always and everywhere, but in fact it was only a shell protecting his real self. He probably was most at ease when he could be gay without feeling that he was losing his prestige, but he felt that this was not the way to behave in one's daily work. It would not be suitable for a scientist to be cheerful except on very special occasions. Over the years these stern manners diminished somewhat, as they were in fact only assumed and were caused by his rather strange ideas of what constituted suitable behaviour. His rule was (he said so himself) that unless one is unpleasant and stern, one does not have much of a chance in this world. It was therefore also logical that as a rule he never praised anyone's work. He assumed a person would be further urged on and work harder next time, if he were not praised for his previous work.

He had other peculiarities too, as for instance that nobody must know the amount of his fortune. He always gave the appearance of being poor. He put his money in different places in order that no bank or savings department should know how much he owned. As a result there were misunderstandings after his death, until it was found that he had left a considerable fortune according to Danish standards, and this in spite of the fact that he had each year donated financial aid and assistance to many persons. Hansen was in fact extremely charitable, but always in secret. He never forgot his poor youth, and he knew better than most how much financial assistance at the right time could mean. A large part of his fortune was left to a Foundation which was to bear his name. The interest on the amount was to be used for an annual prize for work in the biological sciences, the recipients to be chosen by an international committee.

His hard youth had made Hansen a bitter man, and he often regarded life far more pessimistically than justified. Thus he would often imagine himself being attacked without actual reason. He was extremely ambitious and appreciated deeply all honors and distinctions, so it was a great pleasure for him that he did receive so much recognition. He felt that he had been too retiring and not sociable enough, too much of a »loner« as he said. »It was my own fault; don't behave like I have, it doesn't pay«, he would say. Even when he died, he was a lonely man, far away from wife and son.

He is buried at »Western Churchyard« close to the Carlsberg Laboratory, a place where he spent many hours walking about, watching the first spring flowers, the mushroom flora in the fall, listening to the song-birds nesting there, in short, watching all that a true nature lover delights in. His wife is resting next to him; she died a month after him.

While writing this biography I have been guided by one of my old teacher Hansen's principles: »When writing about a person, don't use empty, panegyrical, enthusiastic words, don't use high praise which means nothing. Let his deeds speak for themselves. Explain his work in such a way that his importance is obvious, and explain his character so he appears as the person he really is. Then those who know how to judge will draw the right conclusions.«

Albert Klöcker

1862–1923

by *H. Schiönning**

Albert Klöcker was born June 28, 1862 in Copenhagen, the son of accountant and cashier Ludvig Klöcker, and his wife, née Hahn. He finished grammar school in 1880 and began to study medicine but gave it up after a few years and switched over to pharmacy. In 1888 he took his final examination and thereafter held jobs in various pharmacies in Copenhagen.

On February 1, 1892 he obtained the position as assistant at the Carlsberg Laboratory's Physiological Department, directed by professor Emil Chr. Hansen. Here Klöcker worked for 31 years, first as assistant and after Hansen's death in 1909, as Director Extraordinary until he died in 1923.

Klöcker was well trained for his work at the Carlsberg Laboratory. Already in his youth he was deeply interested in zoology and botany to which he devoted himself in his spare time. In view of his inclination to science and his thorough knowledge of botany, it seemed only natural that he applied for the vacant post as Hansen's assistant. After years of opposition and struggle, Hansen's fundamental investigations and theories had finally been acknowledged throughout the various branches of the yeast industry. While his scientific work had won general approval, there remained much to be done in the new field; much confirmation and amplification was needed, and some early mistakes had to be corrected. For all this Klöcker proved to be a capable and deeply interested collaborator.

Klöcker had versatile and critical faculties which enabled

* *Written in 1924.*

46 A. Klöcker (1907).

him swiftly to become acquainted with a new problem and to penetrate to its core. Moreover, he had a natural talent for and great dedication to the exact sciences. Thanks to these characteristics, combined with an observant eye and manual dexterity in the laboratory, he was eminently suited for his position. Besides assisting in Hansen's work, he also found time to produce a series of valuable independent investigations.

After Hansen's death in 1909 Klöcker was made Director Extraordinary in the same department. It was no easy task that he assumed. Hansen had been the pioneer and founder of his own limited field of yeast physiology. He had established the main lines and decisive principles and it would therefore be difficult to open up new ground. Klöcker's task was more or less staked out as a continuation of Hansen's initiative, and amplification and conclusion of his work. From 1909 to 1914, Klöcker performed this work in the main building of the Carlsberg Laboratory, but then, after the death of H. Schjerning (who had been Director Extraordinary affiliated with the Chemical Department), he was able to move to Schjerning's premises in the old laboratory, situated in the brewery.

Over the years a considerable number of disciples and students, especially foreigners, had frequented the Physiological Department of the Carlsberg Laboratory in order to get acquainted with Hansen's techniques and practical results. Klöcker participated in superintending these courses and he always knew how to arouse the students' interest and how to convey knowledge.

Those acquainted with Klöcker will remember him as a pleasant man who looked upon life without undue gravity, endowed with wit and a full sense of humour. This held true throughout the best years of his life, before illness (a diabetes) taxed his humour and energies and diminished his working capacity.

He died October 3, 1923 after a very short illness. He left a wife, née Ravn, and two daughters.

Research in the field of yeast physiology was Klöcker's main occupation throughout life, but he had many other interests too. His love for botany and zoology never abated, and entomology especially captivated his interest. He loved nature, was an ardent collector and had probably his happiest hours of leisure wandering in the open with a collecting-net in his hand. He made noteworthy contributions to entomology, e.g. it was he who prepared the volumes on butterflies and moths for the »Danmarks Fauna«. From 1903 to 1913 he was also editor of the Danish Journal »Entomologiske Meddelelser«. Klöcker found writing easy and liked to engage in journalistic activities. He was thus first a correspondent, later (1894–1911) an editor of the »Archiv for Pharmaci og Chemi«, and from 1909 to his death he edited the periodical »Bryggeritidende« (Journal of Brewing). Moreover, he was from 1913 a co-editor of »Zentralblatt für Bakteriologie«.

The following is a short outline of the main features of Klöcker's investigations in yeast physiology.

Toward the end of the last century the question of pleomorphism in the Saccharomycetes cropped up repeatedly, as there were those who believed that the Saccharomycetes originated from moulds. Klöcker contributed decisively to the rebuttal of this contention by means of a series of exact experiments which decided the question once and for all.

It was only natural that Klöcker's observant eye took notice of new forms of yeasts, when such forms occurred in the numerous samples of fermentation organisms that were analyzed in the laboratory over the years. Knowledge of such new forms is of greatest importance for taxonomy. Klöcker carefully described a number of new species and members of new genera in the family of Saccharomycetes and in addition new species of Penicillium and Endomyces which were of special taxonomical interest. In the course of his investigations of 17 »apiculatus« forms, he also found new species by showing that the group that had been included under the designation »Saccharomyces apiculatus« is in reality composed of two groups, each consisting of several species; one of these groups belongs to the family Torulaceae, the other to the Saccharomycetes.

Another field full of uncertainties and contradictions was the relation of alcohol producing yeasts to the various sugars. Klöcker performed extensive investigations of the behaviour of various yeasts toward a number of sugars, and he could show that the relations found are constant and can be used to characterize the species of yeasts. As a tool for these investigations he developed a method for the determination of small quantities of alcohol in fermenting fluids.

Klöcker's name became widely known through his excellent book »Fermentation organisms« which appeared in many editions and several languages. This proved to be a most influential work; the author's practical and critical faculties enabled him to create a manual of considerable usefulness for laboratory work with fermentation organisms, thanks to, inter alia, the care taken in discussing apparatus and technique. The book has been widely read and appreciated and will form a lasting memorial to his name.

In addition Klöcker contributed sections on fermentation organisms to several manuals, and many review articles to various journals.

Johannes Schmidt

1877–1933

by *Vilh. Ege**

On February 21st 1933 Professor Johannes Schmidt of the Carlsberg Laboratory died at the early age of 56, after a short illness.

Not only biological science in Denmark, but also international biological research has lost one of its most original and far-seeing workers.

Ernst Johannes Schmidt was born on January 2nd 1877 at Jaegerspris; his parents were the estate inspector Ernst Schmidt and his wife Camilla (née Kjeldahl). He matriculated in 1894 and only four years later took his degree in natural history, with botany as chief subject. In spite of his youth he was able immediately after his examination to undertake independent scientific research, and published in 1899 a small paper on a plant-anatomical subject and a large monograph on the Cyanophyceae of Denmark. This last investigation led Schmidt to work on bacteria, and in 1899–1901 he and Fr. Weis published »Bacteria. Natural history basis of bacteriological studies«**, of which Schmidt himself prepared the first section, dealing with morphology and development, and the third systematic section.

In 1899 Schmidt travelled to Siam with the zoologist Dr. Th. Mortensen in order to carry out a thorough botanical investigation of the little island Koh Chang in the Gulf of Siam. He returned with extensive material, which was partly used in his own investigations and partly treated by other Danish and fo-

**Written in 1933.* ***Translated from the Danish title.*

reign scientists. In 1901 he began to publish »Flora of Koh Chang. Contributions to the knowledge of the vegetation in the Gulf of Siam«, which is an important contribution to our knowledge of the flora of Southeastern Asia. In addition, Schmidt brought back material for a morphological and biological investigation of sprouts and the position of the leaves of the mangrove trees. This investigation was concluded in 1903 with the thesis »Contributions to the knowledge of the sprouts of mangrove trees of the old world«*, for which he was awarded the degree of doctor of philosophy. In the same year (1903) he married Ingeborg, daughter of the brewer S. A. v.d. Aa Kühle. The marriage remained childless.

After the expedition to Siam Schmidt had also begun to work on marine biology and from 1901 he worked with C. G. Johs. Petersen, first as assistant at the Danish Biological Station, and later in connection with the International Council for the Exploration of the Sea. Petersen was one of the leaders of the Danish section of this institution, and in 1903 he sent Schmidt to the Faroes and Iceland with the research ship »Thor«. Schmidt was also during the years 1902–09 an assistant at the University Botanical Laboratory and the Microscopical Laboratory of the Technical University, but during this period his interest was centred chiefly on problems of marine biology, and he produced a long series of papers in this field. The first of these contains well documented accounts of the development of various fish of practical importance from the larval and post-larval stages. These investigations paved the way for the later fundamental biological papers, of which the great 1909 paper »The distribution of the pelagic fry and the spawning regions of the Gadoids in the North Atlantic from Iceland to Spain« has been of widespread importance both for scientific marine biology and practical fishery.

It was during one of these voyages to Iceland (in May 1904)

Translated from the Danish title.

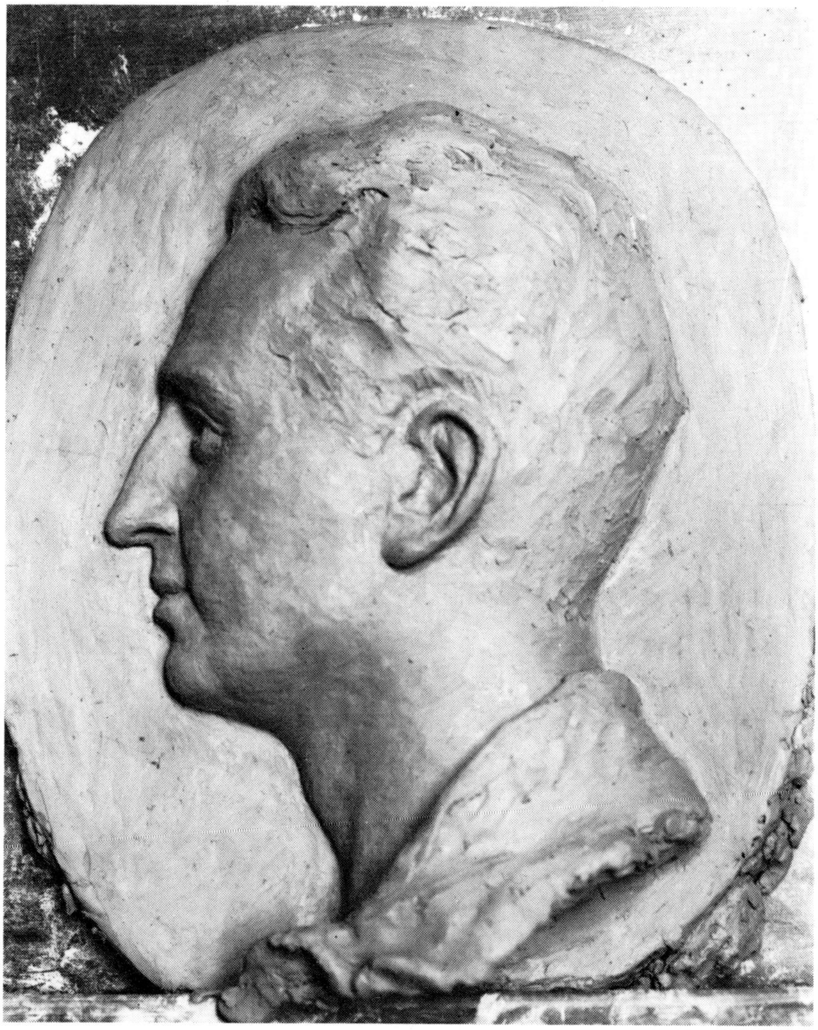

47 Johs. Schmidt (1916). Original clay model for the marble medallion at the Carlsberg Laboratory.

that the event took place which was to prove of such great consequence for Schmidt's contribution to science: the discovery of the first Atlantic eel larva, *Leptocephalus brevirostris,* in the deep sea to the west of the Faroes. The Leptocephalus had previously been found in the Straits of Messina by Italian zoologists, who had shown that this strange organism is the larva of the eel, but nothing was known of the propagation of the north European eel. In the next two years Schmidt succeeded in catching a considerable number of eel larvae in the region between Iceland and the north coast of Spain, and he was able to show that the eels in northern Europe came from the Atlantic ocean, and change to elvers after crossing the coastal banks. However, he found no very young larvae to indicate more accurately the position of the actual breeding places. Schmidt now suspended temporarily his investigations of the open Atlantic, as he thought that at this stage there were better prospects of progress by a systematic investigation of the Mediterranean, where the larvae were first found and where the breeding places were presumably situated. With the support of the Carlsberg Foundation he organised two expeditions to the Mediterranean, in the winter of 1908–09 and in the summer of 1910. These investigations led to the surprising result that the eel does not breed in the Mediterranean.

On his return from this last expedition, Schmidt was met by new problems, as he had been previously selected to succeed Emil Chr. Hansen as head of the Physiological Department of the Carlsberg Laboratory. This appointment induced Schmidt to resume for a number of years his botanical studies.

Schmidt embarked on an experimental investigation of hops, and by extensive experiments on cultivation and crossing he obtained results of both theoretical and practical importance. In the first two papers of the series »Investigations on Hops«, he deals with points of plant physiology, e.g. the longitudinal growth and rotational movement of hop stems, and reports several interesting observations. Schmidt's plan of work developed further in the direction of applied science, and he attacked the problem of producing a hop plant which should combine the high lupulin content of foreign sorts with the hardiness of the Danish hops.

From 1911 onwards he cultivated in the Carlsberg research gardens a selection of hop plants of the most important Danish

48 Harvest of hops (1912) in the greenhouse established by Schmidt for hop breeding at the Carlsberg Laboratory.

and foreign sorts. He carried out investigations on the constancy of the valuable qualities in older »clones«, and also carried out crossing experiments with a view to the formation of new clones which combine various valuable qualities. (The word »clone« is used here to indicate all those hop plants derived from the same seedling by vegetative propagation). In a rough estimate of the quality of hops as a commercial product, the most important points are lupulin content and aroma, and on Schmidt's initiative a quantitative method was developed for estimating the first of these properties. It is not possible to obtain a similar method of ready application for estimating the aroma, since a valuation of this property depends entirely on subjective judgement. Schmidt succeeded however in demonstrating the constancy of the aroma in older clones, and that it was possible to transfer it to the offspring of a cross between foreign and Danish plants. As regards the lupulin content in plants raised by crossing, Schmidt found on crossing wild male plants (supposed to be genotypically low in lupulin content) with foreign female plants that although the average lupulin content of the offspring was less than that of the mother plant, some variants were formed having a greater lupulin content than the mother plant. This discovery opened up possibilities for improving the species, since by vegetative propagation of the offspring rich in lupulin, this valuable property could be retained. Besides the simple ability to yield a good number and weight of hops per bine, the time of flowering of the hop plants is also of importance. In the Danish climate it is important that flowering shall take place early, so as to ensure that the hops may ripen even in bad summers. Just as before, crossing experiments showed that a few variants were formed with a specially early time of flowering, and thus indicating the same possibility of improving the species. By following out these indications Schmidt succeeded in producing new clones suitable for practical use.

Simultaneously with his investigations of hops, Schmidt continued his studies of the eel. Norwegian investigations in 1910 had shown quite clearly that the breeding place of the eel must be far removed from European coasts. Now Schmidt established a collection of pelagic animals by securing the help of ships

49 Schmidt on board the »Dana« II (1922) with a ring-net of his own construction.

crossing the Atlantic. In 1913 he sent the little schooner »Margrethe« on a transatlantic expedition, during which a large number of young eel larvae were caught in the Sargasso Sea. At this point Schmidt realized the necessity of a study of the racial characteristics of the eel before the investigation of its biology could be profitably continued. He investigated by the statistical method of variations a large number of specimens of the European eel *(Anguilla vulgaris)* from very different localities, with the result that it was not possible to detect the presence of any racial differences. It was next desirable to carry out a corresponding investigation with another species of fish having approximately the same numerical features, differing as much as possible from a biological point of view. For this purpose the common blenny *(Zoarces viviparus)* was chosen; since it is viviparous and lacks the pelagic stage, it might be expected to have a tendency to form local races. This expectation was completely confirmed. The difference must therefore be due to the fact that European eels are all of the same origin, while the blenny consists of different and distinct stocks.

This comparative investigation gave Schmidt the idea that the blenny would be very suitable for the study of more general questions, and led him to commence a new series of investigations. This work was carried out at the Carlsberg Laboratory side by side with the investigations on hops. The results were collected under the common title »Racial Investigations«. The first investigation consisted in population analyses based on material from 61 different stations (about 15,000 specimens in all). The characters investigated were the number of vertebrae, pectoral fin rays, hard rays and pigment spots of the dorsal fin. The average values for these characteristics may be combined in very different ways in the different populations. In all the fjords studied it was found that the average value for three of the characteristics mentioned is lowest at the base of the fjord and increases regularly as we pass outwards. Schmidt also showed by making individual offspring analyses that the populations can be subdivided into yet smaller elements, and came to the conclusion that a »Race« is a mixture of different genotypes, and is determined primarily by the quantitative proportion between these genotypes and only secondarily by the environment.

In a subsequent series of papers Schmidt investigated the relation between the genotypical constitution, determined by heredity, and the phaenotypical character, partially determined by the environment. Two other species, the »million fish«, *Lebistes reticulatus,* and the common trout, *Salmo trutta,* were included in the investigations and used for experiments. Investigations were carried out with *Lebistes* to determine the effect of altering one factor in the environment, the temperature. It was found that although the number of organs (rays in the dorsal fin) could be much affected by the environment, there were also differences of an hereditary nature between the different individuals. The same result was obtained in some transplantation experiments with the blenny, in which a large number of specimens were transferred from two different localities to some big wooden boxes immersed at the same place. The change of environment displaced the values for the offspring of the transplanted samples, but the differences between the two populations did not disappear in spite of the fact that their development took place under the same external conditions.

With regard to natural environment, Schmidt advanced many considerations as to which factor or factors were operative in producing racial characteristics, and also as to whether the action was direct or indirect. In this connection he obtained an important result by determining the age of the individual members of some blenny population samples, thus enabling him to separate the different years and obtain the average values of the characters in each year. By this means he demonstrated the occurence of annual fluctuations, which had a parallel course for two widely separated populations, thus showing that the effective factor or factors could not be of local origin.

Some years later Schmidt subjected the Atlantic cod *(Gadus callarias)* to a racial investigation, and was responsible for yet another advance in this field. He investigated ca. 20,000 specimens from 114 stations, and found (besides several points of special interest for fish biology) that in open water there is a correlation between the average values of certain parameters and the temperature. This was especially the case for one of the characteristics investigated, the number of vertebrae, which was found to increase with decreasing temperature.

The central point of Schmidt's interest in these investigations was always the relation between environment and genotype, and in 1930, immediately after the conclusion of the work on the cod, he realized a project he had long entertained: transplantation from one locality to another under natural conditions. Among the localities in which the blenny was subject to a regular annual investigation was a small stretch of water cut off from the open sea, Kjelds Nor at the southern extremity of Langeland. This locality was of special interest partly because of its isolation, and partly because it contained a population which was in several respects peculiar, having among other things the highest number of vertebrae found. During several years of statistical investigations the population at Langholm at the base of Roskilde Fjord had proved to represent the opposite extreme with regard to the number of vertebrae. A transplantation from this locality to Kjelds Nor was therefore considered to be specially favourable, and about two thousand individuals were actually transferred. An investigation of the embryo material from Kjelds Nor in the two succeeding years gave very promising results; however, Schmidt died before he could pursue these experiments.

The question of the effect of external conditions naturally led Schmidt to practical investigations of heredity, to which he contributed the important paper »The value of the individual for breeding as determined by the method of diallel crossings«. This method supplements the so-called method of pure lines of W. Johannsen, which makes possible the determination of the breeding value of self-pollinating plants. A similar determination can be carried out for plants or animals with cross fertilisation by means of Schmidt's method of diallel crossings. In this method each male is paired with each female, and the offspring is investigated after being raised under uniform conditions. From the average values thus obtained it is possible to calculate the differences in the breeding values for the males and for the females. If care is taken to make the figures comparable for all the parents by referring them to the same external conditions, an expression is obtained for the breeding values themselves.

In addition to the activities described above, Schmidt continued his oceanographic investigations, and pursued especially the problem of the biology of the eel. In 1920 he carried out a transatlantic expedition in the motor schooner »Dana«, and in a systematic exploration of the Sargasso Sea he collected such a

50 Johs. Schmidt and his collaborators the botanist Ove Paulsen (left) and the hydrographer J. N. Nielsen (right). (»Dana« II at Gibraltar 1928.)

representative selection of eel larvae that he was able in the main to fix the breeding places of both the European and the American eel. He demonstrated the surprising fact that the breeding places of the two Atlantic species are partly overlapping. It was also possible with the help of earlier material to calculate the respective periods of development for the larvae, so that the essential points in the biology of the Atlantic eels were thus cleared up. However, Schmidt wished for two reasons to carry out a general oceanographic investigation of the North Atlantic; in the first place to do a comparative study of other organisms in order to obtain a better understanding of the contrasts between their biology and that of the eel, and in the second place to solve the wide range of problems arising out of the numerous preliminary results obtained by the working up of the material of the two Mediterranean expeditions.

51 Reception at the Carlsberg Laboratory on the occasion of the International Council for the Exploration of the Sea meeting in Copenhagen March 1931 (names of persons see chapter 14).

Before long Schmidt had obtained the necessary support for such an investigation, and in August 1921 he started on what was then his longest expedition, lasting eleven months. This time he had at his disposal a specially equipped vessel, the research ship »Dana« and a large scientific staff. The expedition returned with very valuable biological and hydrographical material, which has amply fulfilled Schmidt's double purpose and has served to place Danish oceanographic research in the first rank.

Schmidt was not, however, satisfied with the investigation of the North Atlantic. Just as the biological exploration of the Mediterranean could not be brought to a satisfactory conclusion without supplementary material from the Atlantic, so also workers on the latter material would often feel the want of comparative data on the state of affairs in other oceans.

52 On the occasion of the »Dana«s return an exhibition of ethnographical objects was held at the Carlsberg Laboratory, some of them shown by the laboratory's assistants and friends (1931).

Schmidt himself was acutely conscious of this lack in the case of the data on the eel. He had found that both the Atlantic species can only breed in deep tropical waters, and had proposed the hypothesis that this was due to the fact that the genus itself was of tropical origin. He had, however, no proof of this and he realized that before further progress could be made it was first necessary to carry out a thorough revision of the taxonomy of the *Anguilla* genus and the geographical distribution of the various species, current knowledge of which left much to be desired, owing partly to the extremely intricate mutual relationships of the species.

Schmidt was of the opinion that such an investigation of a ubiquitous genus would also be very valuable for the racial investigations, and could only be carried out by the same methods as used for the Atlantic eel. He therefore incorporated it in the work programme of the Carlsberg Laboratory; in the course of several years practically all the specimens of *Anguilla* in the museums of the world were examined, and a very large number of samples suitable for variation-statistical treatment were obtained from all parts of the world where the genus is extant.

In 1927 these investigations were so far advanced that Schmidt was able to lay plans for the great expedition which was specially intended to investigate the biology of the *Anguilla* species of the Pacific and Indian oceans, but which was also to carry out general oceanographic investigations. A handsome grant from the Carlsberg Foundation made it possible to carry out this plan, and in June 1928 »The Carlsberg Foundation's Oceanographical Expedition round the World 1928–30« set out. The expedition returned with extremely extensive hydrographical and biological material, which in the case of the pelagic organisms was by far the greatest that any single expedition has hitherto produced. The collections were carried out according to the same biological and bio-geographical principles as used in Schmidt's earlier expeditions, a sufficient guarantee that when studied in Schmidt's spirit they will very greatly increase our knowledge of the animal life of the ocean.

Schmidt was only in part able to see the fruits of his last great venture, but his pupils will receive the heritage and will do their best to give the Carlsberg Foundation's expedition a place in the history of oceanographic exploration which will be a worthy memorial to Ernst Johannes Schmidt.

Johannes Schmidt's expeditions.

*1899–1900		Siam
*1903	»Thor«	Iceland-Faroes.
*1904	»Thor«	Iceland-Faroes.
*1905	»Thor«	North Sea, Iceland-Faroes, W. of British Isles, Bay of Biscay.
*1906	»Thor«	North Sea, Channel, Bay of Biscay, W. of British Isles.
*1907	»Thor«	Skagerak.
*1908	»Thor«	North Sea, Iceland-Faroes, W. of British Isles, Mediterranean.
*1909	»Thor«	Mediterranean, Atlantic west of Europe. – Hired vessel: Iceland.
*1910	»Thor«	Atlantic west of Europe, Mediterranean. – Hired vessel: Faroes.
1911	»Thor«	Collections in the Straits of Messina (by Capt. G. Hansen).
1912		Hired vessel: Faroes.
*1913	»Margrethe«	Faroes, North Atlantic, West Indies.
1911–1915		Various ships collecting in the Atlantic, Mediterranean, Red Sea.
*1920	»Dana« I.	North Atlantic.
1921	»Dana« I.	Sargasso-Sea.
*1921–1922	»Dana« II.	Mediterranean, North Atlantic, Caribbean Sea, Gulf of Panama.
1923		Hired vessel: Faroes.
*1924	»Dana« II.	Faroes-Iceland.
*1925	»Dana« II.	Faroes-Iceland. Greenland.
1926	»Dana« II.	Faroes-Iceland.
1927	»Dana« II.	Faroes-Iceland.
*1928–1930	»Dana« II.	The Carlsberg Foundation's oceanographical expedition round the world.
1931	»Dana« II.	Faroes-Iceland, East Greenland, North Atlantic south to Madeira and the Azores.
1932	»Dana« II.	Faroes-Iceland, East Greenland.

* Professor Schmidt himself partaking in the expedition, in the whole expedition or part of it.

Øjvind Winge

1886–1964

by *M. Westergaard**

Øjvind Winge was born in Århus (Jutland) on May 19th, 1886; his father, Sigfred Victor Winge, was a prosperous lawyer. The family can trace its roots back to the 17th century and the zoologist Herluf Winge was a distant relative. His mother, Agno Rian, came from Trondheim in Norway.

Winge came from a social environment which – not without justification – considered the study of natural history a certain road to financial ruin and social disgrace. After graduating from Marselisborg grammar school in 1905 he therefore spent his first months at the Copenhagen University studying law, thereby fulfilling a promise to his parents. However, his interest in natural science was too deeply rooted and he soon switched to the Faculty of Science, where he studied natural history (botany, zoology, geology and geography) with botany as the major subject.

Winge's interest in biology goes back to his boyhood, and he owned his own microscope at the age of fifteen. During his highschool years he became acquainted with an eminent mycologist, C. Ferdinandsen, with whom Winge formed a lifelong friendship and with whom he established a close scientific co-operation over many years. Also Poul Larsen, the micromycete specialist, took Winge on mycological excursions.

It is therefore no wonder that Winge specialized in mycology during his studies in Copenhagen. His teachers here were Eugenius Warming, the Botany Professor, L. Kolderup Rosenvinge, the Professor of Cryptogamic Botany, and Wilhelm

**Written in 1964.*

53 Øjvind Winge 1956.

Johannsen who was Professor of Plant Physiology. Also the mycologist E. Rostrup helped him in his mycological studies. During the excursions with the Danish mycological society he became acquainted with Emil Christian Hansen who was professor at the Carlsberg Laboratory and who had developed the technique of maintaining pure cultures of yeast. Emil Chr. Hansen offered Winge a position at the Carlsberg Laboratory on the condition that he studied plant physiology instead of mycology – an offer which he wisely declined.

When Winge graduated from the University in 1910 with a »mag. scient« degree in natural history (botany), he had already published six papers, all in collaboration with Ferdinandsen, and all on mycological subjects.

The Danish mag. scient. degree is more or less equivalent to a Ph.D. degree, and Winge prepared for his post-graduate work by studying cytology at various laboratories. His travellings, which were financed by a grant from the Carlsberg Foundation, brought him first to Otto Rosenberg and G. Lagerheim at Stockholms Högskola. The winter of 1910–11 was spent at the Sorbonne with professor P.-A.-C. Dangeard, and the rest of the grant was used for a visit to the United States where he worked in the laboratory of professor J. M. Coulter and C. J. Chamberlain in Chicago.

In his grant application to the Carlsberg Foundation, Winge stresses the importance which chromosome cytology may have for his mycological studies, but he also mentions that the study of chromosome cytology is of importance for the understanding of phenomena of heredity. Nevertheless, there is no doubt that when he went abroad, his outlook was that of a botanist and mycologist, he had not yet become a geneticist. The immediate results of Winge's newly gained knowledge was a paper on the cytology of *Plasmodiophoraceae*.

On his return to Copenhagen, Winge was appointed research assistant to professor Johannes Schmidt, Emil Christian Hansen's successor at the Physiological Department of the Carlsberg Laboratory. Although originally trained as a botanist, Schmidt was to gain fame as an oceanographer. However, Schmidt was also deeply interested in genetics. There is no doubt that Schmidt had a great influence on Winge's transformation from a mycologist and cytologist to a geneticist during these early years at the Carlsberg Laboratory.

Schmidt operated his laboratory according to the old European tradition which meant that during the official working hours the scientific staff were the professor's personal assistants, and whatever work they wanted to do on their own had to be done in their spare time. This, however, was a great improvement over Emil Christian Hansen's habits: he simply locked the laboratory at 5. p.m. – and he had no pupils.

As Schmidt's assistant Winge indeed became a »jack of all trades«. Among other things he developed a method for determining resins in hops, and he invented a new heating device for aquaria. Winge once told me that his most agonizing task as Schmidt's assistant was to register the ocean currents. This was done by throwing Champagne bottles with messages overboard from various ships. Winge had to fit piles and piles of – alas empty – bottles with message cards, and sometimes he wished he had chosen the more glamourous career of an oceanographer. Sufficient numbers of cards were recovered and mailed to the Carlsberg Laboratory.

However, in Schmidt's department Winge also became acquainted with two organisms with which he was to work during most of his lifetime. One was *Humulus* (hop). His first *Humulus* paper was published in 1914. His last paper which was published a few months before his death was a survey of his 50 years of breeding work with hops. The other species was the small aquarium fish *Lebistes reticulatus* which was to become the subject of some of his most important genetical work.

In his spare time Winge collected material for his doctor's thesis which was published in 1917, under the title: »The chromosomes. Their numbers and general importance.« Winge's thesis is, as it had to be, a rather heterogenous paper, but it contains a hypothesis which has become one of the fundamental theories in cytogenetics, viz. the theory of speciation in plants through polyploid evolution by means of species hybridization followed by doubling of the chromosome number of the hybrid.

During the first decades of this century a number of cytologists had observed that the chromosome numbers of related plant species often form arithmetical series, now called polyployd series. To these series Winge contributed with own counts, especially of *Chenopodiaceae*. He also made the first statistical survey of the distribution of plant chromosome numbers, a survey which left no doubt as to the occurrence of polyploid series. In a related field, O. Rosenberg in Stockholm (with whom Winge had studied) had in 1909 published the first investigation of meiosis in a sterile species hybrid, *Drosera longifolia* × *rotundifolia*. Furthermore, in 1913 Harry Federley, the Finnish zoologist, had studied meiosis in sterile butterfly species hybrids, belonging to the genus *Pygaera*. The work of these two cytologists indicated that one, perhaps the predominant, reason for the sterility of species hybrids is that they are unable to go through a normal meiosis, because homologous chromosome pairing at pachytene is not possible. Winge now proposed the hypothesis that chromosome doubling of such hybrids would solve the homology problem and make normal meiotic chromosome pairing possible. He therefore predicted that such chromosome doubled hybrids would be fertile, breed true, and in all essential respects behave as new species. Hybridization followed by chromosome doubling would thus explain polyploid evolution.

This was a brilliant and purely theoretical deduction which is typical of Winge's ability to combine facts and observations from related fields and formulate a simple hypothesis which lends itself easily to experimental verification.

Although Winge was wrong in minor details – as to the actual mechanism of chromosome doubling – his main hypothesis was soon verified by the discovery of various amphidiploids (the term was not coined by Winge, but later by the Russian geneticist S. Navashin), like *Nicotiana »digluta«*, *Pri-*

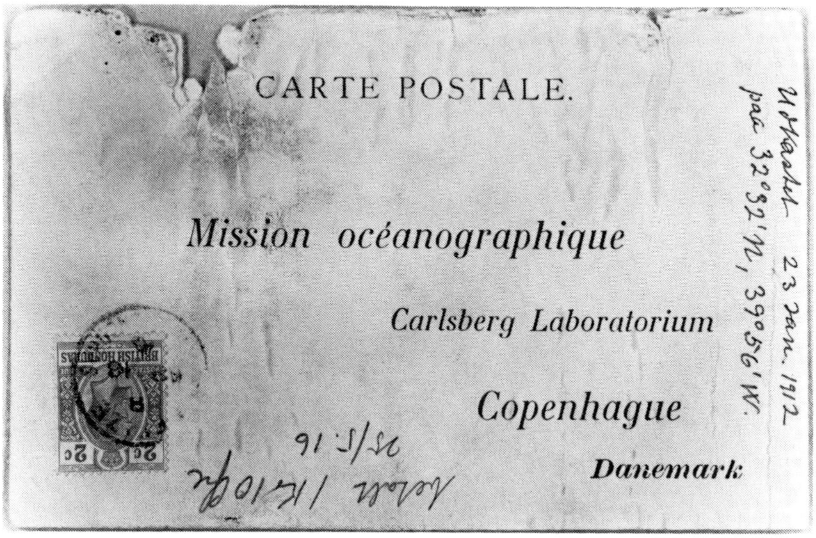

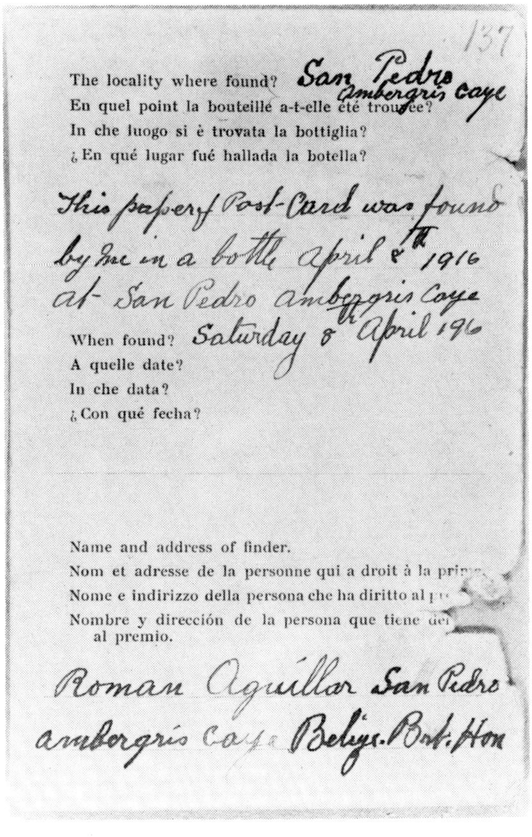

54 One of the Champagne bottle messages.

55 Winge as Johs. Schmidt's assistant (1913).

mula »Kewensis«, etc. etc., and the study of polyploid evolution was for many years – and as a matter of fact still is – one of the most fruitful fields of cyto-genetics and of cyto-taxonomy.

Winge always took a great interest in species hybrids. He published papers on *Tragopogon* hybrids, *Geum* hybrids, *Sambucus* hybrids, and even animal species hybrids. He also described, in collaboration with H. N. Frandsen, the amphidiploid *Brassica napocampestris*. His excellent investigations on micro-speciation through polyploidy in the plant genus *Erophila* should also be mentioned.

Despite the many investigations mentioned above, Winge did, however, not make any further fundamental contributions to the study of polyploid evolution after the publication of his 1917 paper. This was typical of Winge's method of work, especially in his younger years. He liked to skim the cream of a problem, but would then switch to something else, and leave the following-up to others. It was this method of work – running contrary to the general trend among geneticists who were generally inclined to specialize on certain organisms and on certain subjects – which made it possible for Winge to work on such a variety of organisms and on such a variety of problems.

In the years 1919–23 Winge published, among other papers, a clearcut, but still puzzling case of cytoplasmic inheritance of variegation in *Humulus japonicus,* and an investigation of the inheritance of hair colour in horses. In this period he also made a study of the inheritance of eye-colour in man, and his first papers on *Lebistes,* including the first case of X-Y-linked inheritance came out in 1922/23. He also published his first – and only – paper on *Drosophila*; the subject of the paper was a balanced lethal system. This was his only attempt to »join a bandwagon«.

Winge's thesis and his cytogenetic investigations had an important impact upon Danish biology. Like several others of the Founding Fathers of Genetics, Wilhelm Johannsen, by then in his sixties, for many years did not accept the gene-chromosome theory, and to Winge is due the credit of having introduced chromosome cytology and cytogenetics into Denmark. After his work on pure lines, Johannsen »rested on his laurels« – to use his own expression – and the new impulses to Danish genetics came from Winge's laboratory. Although he was not a pupil of Johannsen, Winge had a great admiration for Johannsen's genius. This is expressed in Winge's later biography of the great Danish plant physiologist and geneticist.

In 1921 Winge was appointed to the newly established chair in genetics at the Veterinary and Agricultural University in Copenhagen. This appointment provided him with excellent research facilities because he had, in addition to the laboratory in Copenhagen, a good experimental field in Lyngby, eight miles from Copenhagen. Winge took full advantage of these facilities, and assisted by a number of able young scientists: Jens Clausen, Aage Skovsted, C. Syrach-Larsen, Esben Ditlevsen and Louis Harmsen, he accomplished an incredible amount of work. Winge was undoubtedly, and especially in his early years, a hard master to serve. He was, as already mentioned, brought up in the tough traditions of Johs. Schmidt and Emil Christian Hansen, and only the fittest could survive scientifically. The story is still told with awe among Danish scientists of how Jens Clausen (who later became a member of the Carnegie Institution at Stanford) bicycled – often accompanied by his wife – the eight miles out to Lyngby at 5. a.m. There he would work on his violets until 9. a.m. from which time he would work on Winge's material until 4. p.m., when he could return to the violets, until he bicycled home at 11. p.m. The modern way of treating budding scientists is undoubtedly more human, but certainly also less selective. It should also be put on record that Winge and his assistants could not have accomplished what they did at the Agricultural University without the devoted co-operation of his gardener Mr. Otto Mathiesen whose capacity and enthusiasm for hard work fully matched their own.

In order to fully appreciate Winge's achievements as professor at the Agricultural University it must be realized that during all this time he continued his mycological work, mostly in collaboration with Ferdinandsen, and that in addition to his many publications in foreign languages, he wrote many papers and several books in Danish.

In the years 1907–30 Ferdinandsen and Winge wrote more than 20 papers on mycology, and their co-operation culminated with the publication, in 1928, of their widely used mycological field handbook (in Danish), of which a second edition appeared in 1943. Most of these papers, including the mycological flora, are adorned by Winge's excellent drawings. He was a very skilled artist, and in his younger years he did a great deal of oil painting of landscapes in his spare time.

In 1928 Winge published, in Danish, his Textbook in Genetics which appeared in three editions (1937 and 1943). In this textbook is included a great deal of Winge's otherwise unpublished material. The textbook was widely used and for several decades it was the most popular genetical textbook, not only in Denmark, but also in the other Scandinavian countries.

Winge's most important work from that period is undoubtedly on the cytogenetics of sex determination. In 1923 he described some of the first cases of heteromorphic sex chromosomes in plants. I believe that Winge, for priority reasons, published this paper too soon, and some of his observations and interpretations of sex chromosomes in *Melandrium, Humulus* and *Vallisneria* did not stand up to the scrutiny of further investigations. The cytological work on *Melandrium* was followed up by a number of excellent genetical papers on sex-linked inheritance, in which he described the first cases of Y-linked and X-Y-linked inheritance in plants, and on sex determination.

Parallel with his botanical studies the important work on *Lebistes* was continued. In this species he extended his earlier observations on various types of inheritance of colour genes. He also showed that by artificial selection it is possible to produce strains with female heterogamety from a population which originally showed male heterogamety.

In his interpretation of the genetics of sex determination Winge was close to C. B. Bridges' interpretation of the *Drosophila* work, but he was critical towards R. B. Goldschmidt's theory of sex determination, and showed that Goldschmidt's *Lymantria* work could be interpreted through a conventional balance hypothesis without invoking cytoplasmic factors; this criticism brought him into a heated discussion with Goldschmidt. Winge got into quite a few scientific discussions during his lifetime, including one about the Sargasso Sea! There is no doubt that Winge enjoyed such arguments. He was very critical, always a good fighter, and, unlike most Danes, he would rather fight than compromise – as the trustees of the Carlsberg Laboratory can undoubtedly testify. There was, however, nothing personal about these discussions; I know that he both liked and admired Goldschmidt, and his criticisms were almost always useful and constructive. Only in one case did he – in my opinion – loose his temper: His criticism of yeast cytology (1951), although scientifically sound, seems somewhat arrogant.

In two papers, from 1927 and 1930, Winge described the first cases of hyperploidy and other chromosome abnormalities associated with tumour growth in plants, and animals, respectively. Unfortunately, these investigations which Winge himself ranked among his most important works, were published in Germany in a journal which was not widely known among geneticists from the English speaking world. For many years Winge was grieved that these publications gained little recognition, and he was very pleased when, in recent years, he received full credit as the pioneer in the now so fashionable field of chromosome patho-cytology. It seems appropriate to republish some of his original, now classical, drawings on this occasion (Figure 56).

Winge had excellent research facilities at the Agricultural University, but as the teaching and examination load was rather heavy, he made several attempts to go back to the University of Copenhagen. Wilhelm Johannsen was Professor of Plant Physiology at the University and he also taught the genetics course. During his last years of tenure Johannsen was relieved of his teaching duties in order to prepare the foundation for a Genetics Institute and chair. However, no plans had materialized when he died in 1927.

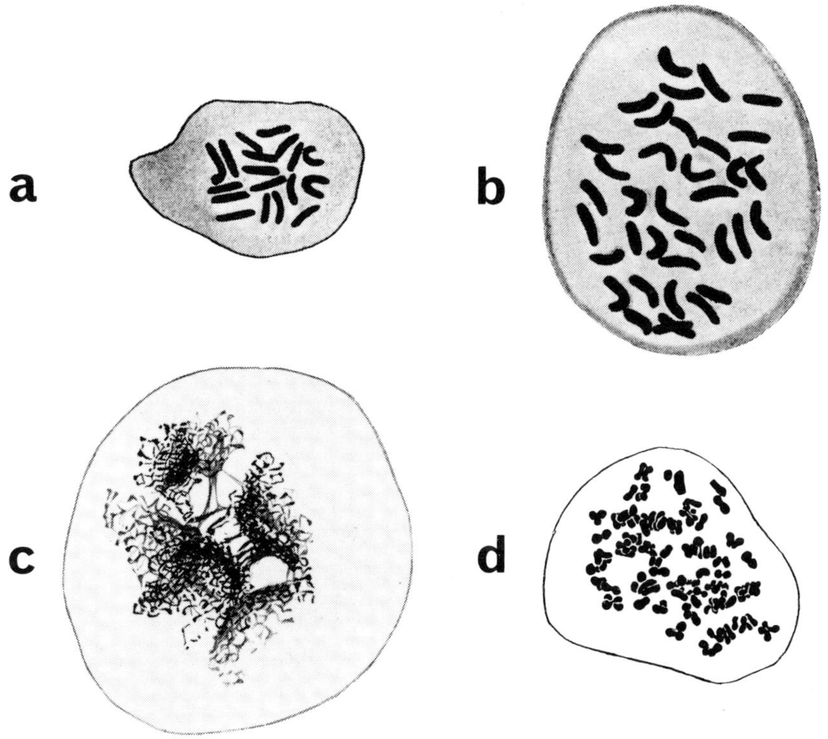

56 Winge's drawings of tumour cells.
a: diploid and b: tetraploid metaphase plate in tumour cells of a »Crown gall« tumour in sugar beets. c: cell with multipolar spindles in a mouse carcinoma. d: polyploid cell in a mouse carcinoma. (Magnifications, see chapter 14.)

From 1929 to 1935 Winge taught the elementary genetics course at the University. It was based on his textbook. He also gave an exellent elementary course in biometry. At this time his good friend, C. H. Ostenfeld, the Professor of Botany, was planning to build an Institute of Genetics and Plant Physiology in the Botanical Garden.

Winge would undoubtedly have liked to succeed Ostenfeld in the chair of Botany. However, he did not want to apply for the professorship and the Faculty did not offer it to him.

In 1933 Johannes Schmidt died, and when the trustees of the Carlsberg Laboratory offered Winge the Directorship of the Physiological Department, he accepted the offer, realizing that the possibilities of his obtaining a professorship at the University were now very remote. Although Winge's appointment to the Carlsberg Laboratory was an excellent choice, both for Winge himself and for the Carlsberg Foundation, the appointment was a serious set-back to genetics at the Copenhagen University. A chair in genetics was not established until 1949, and the institute which Winge, Ostenfeld and Boysen Jensen had planned in the late twenties did not materialize until 1958. Then, with the Rockefeller Foundation again acting as an indispensable catalyst, it became possible to build an institute, now also comprising microbiology and biochemistry in the Botanical Garden*.

The charter of the Carlsberg Laboratory states that »the men in charge of the Laboratory, in addition to their work on the special subject of the Institute, should strive to maintain their scientific ability in general by other studies and investigations, but in such a way that the main task of the Institute is not neglected or forgotten«.

The »main task« of the Physiolopical Department had undoubtedly been somewhat neglected during the later years of Schmidt's directorship when his whole work was devoted to oceanography. Winge, however, felt very strongly about the obligations of the director towards the Charter of the Laboratory and with his usual energy he began to put things straight, immediately after he had taken over. During the rest of his life he worked almost exclusively on the three organisms which are the most important to the brewing industry, viz. hop, barley and yeast. From 1938 on, the hop and barley work was carried out on a farm called »Nordgården« on Sealand. This farm was owned by the Carlsberg Breweries, not by the laboratory, and Winge continued until 1963 to direct this work after his retirement from the laboratory in 1956.

There is no doubt that the yeast work gave Winge, in addi-

* compare Westergaard's biography.

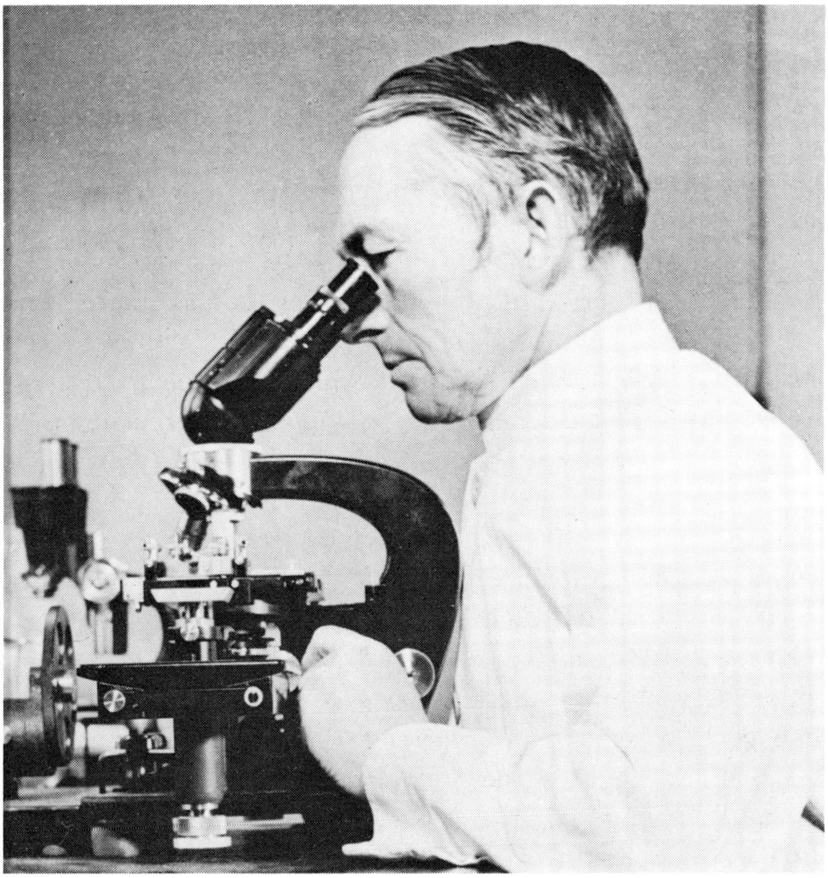

57 Winge working with the micro-manipulator (1942).

tion to new fame, the greatest intellectual satisfaction. He was also very pleased that the new and important break-through in yeast biology – now to be described – came from the Carlsberg Laboratory. He was unusually well qualified to tackle the problems of yeast genetics, because he had behind him 25 years of experience as a mycologist and more than 20 years of experience as a geneticist. For the first time he was now able to combine his two main scientific interests, mycology and cytogenetics in a single project, and his approach to the problem of developing yeast genetics has a touch of sound professionalism which commands the greatest admiration.

First Winge collected his stock cultures, mainly from the old Carlsberg collections of Emil Christian Hansen and Albert Klöcker. Next, he designed new culture chambers and new techniques of micro-manipulation, adapted to the yeast work. The break-through came already in 1935, when he demonstrated for the first time that yeasts reproduce sexually and have a regular haplophase-diplophase cycle. This was established by purely cytological methods. The first genetical work in which Mendelian segregation in yeast was demonstrated followed in 1937. This paper was published in collaboration with O. Laustsen, to whose remarkable technical skill Winge owed much of his early success in the yeast work. During his Carlsberg years, Winge had as research assistants J. H. Wanscher, Aase Hjort, Aage Skovsted, Otto Laustsen, Esben Ditlevsen, Vagn Hartelius, E. Bille Hansen and the American-born geneticist Catherine Roberts.

In 1938, again in collaboration with Laustsen, Winge described the first case of artificial species hybridization in yeast. During these years he also finished a number of investigations which he had begun at the Agricultural University, and his last *Lebistes* paper appeared in 1947. From 1949 on all his work was, however, devoted to yeast genetics. This includes an excellent investigation of a balanced lethal system in *Saccharomycodes Ludwigii;* a somewhat controversial paper on cytoplasmic inheritance in yeast, and the later investigations on the inheritance of fermentation ability in yeast. These last papers, which are all published in collaboration with Catherine Roberts, show various types of gene interaction in control of fermentation ability.

Winge's yeast work is summarized in several important review articles and monographs, published in collaboration with Catherine Roberts and others.

Although Winge was among the first geneticists to work on microorganisms and although he used differences in fermentation ability as genetic markers in his later work, he was not among the founders of microbial, biochemical genetics. One must remember that Winge's basic training, 30 years previously, had been in natural history (botany, zoology, geology and geography). It was undoubtedly a pity that he never collaborated with the »ground floor« of the Carlsberg Laboratory which housed Linderstrøm-Lang's department of chemistry. Had such a collaboration been established, biochemical, microbial genetics might well have been born in the thirties at the Carlsberg Laboratory, and not in the forties at Stanford (G. W. Beadle and E. L. Tatum), and at Uppsala (Niels Fries).

It is impossible to deal with the last part of Winge's yeast work without mentioning his running arguments with the American geneticist Carl Lindegren. In many ways these two brilliant scientists complemented each other. Lindegren: Imaginative, controversial and often leaving a few loose ends to be tied up in his publications. Winge: Critical, solid, thorough and sound, and loosing no time in pointing out the weak points in Lindegren's arguments. This did not mean that Winge was always in the right and Lindegren always in the wrong. In most cases the truth was trashed out between them and to Lindegren goes the undisputed credit of having discovered the mating type system in yeast and »gene conversion«. Winge reluctantly accepted this last phenomenon – although he had a sound aversion against this obnoxious term – and in one of his last papers he proposed a model to explain allelic crossing over. This model, although logical and thought-provoking, also shows Winge's limitation in his later years: He was unable to master the new science of molecular genetics.

When some day the history of yeast genetics will be written, it will no doubt be acknowledged that its rapid progress owes a great deal to the Winge-Lindegren discussions. It should also be recorded that despite their many arguments the two antagonists liked and respected each other, and both recognized in his opponent a scientist of the highest scientific integrity.

Winge was deeply interested in the social implications of genetics, but most of his writings on these subjects are in Danish. He was among the first to warn against the genetic hazards of ionizing radiation, a warning which he repeated a few years ago when the atomic age had made these problems even more important. Although he was fundamentally a shy man who undoubtedly disliked publicity very much, Winge adressed a large public audience of Danish students in 1938. On this occasion he explained, as was his habit, in no uncertain terms his opinion about the Nazi racial theories which were beginning to creep across the Danish border and infect some of the Danish political parties.

Like most Scandinavian geneticists of his generation, Winge was deeply interested in applied genetics. He made several overtures to the animal breeders, but he got a very cold reception, because Winge, very much the city bred aristocrat, simply did not speak their language.

He had, on the other hand, an indirect impact upon Danish forest tree breeding, because C. Syrach-Larsen, the founder of the genetical school of forest tree breeding, was his pupil and had been his assistant at the Agricultural University. Winge always followed Syrach-Larsen's work with great pride and interest.

As a plant breeder, Winge made more direct contributions to Danish agriculture. His barley breeding work, which aimed at the production of a high yielding malt barley, was very successful. His »Carlsberg II« barley, which is still widely grown in Denmark, was in most of the years 1950/57 listed as the highest yielding malt barley in the field trials of the »European Brewery Convention Barley Committee«. It is not Winge's fault, but due to other considerations, that the Danish breweries import their hop, because the hop strains which Winge developed over his 50 years of breeding work are excellent.

Winge's discovery of sexual reproduction in yeast also opened new possibilities for producing better strains through crossing. These methods, which he once called »breeding under the microscope«, have resulted in the production of new strains of Bakers' yeast, but they have not been succesfull in producing better strains of Brewer's yeast.

Although many foreign students came to work with Winge, especially in order to learn his yeast techniques, I do not think one could consider him as the head of a »Genetical School«. It is told that C. F. von Weizsäcker, the German physicist, once said about Niels Bohr, the founder of the famous »Copenhagen School« of physics, that Bohr lacked two qualities which were normally characteristic of a head of a school: He was not a pedagogue and he was not a tyrant. Winge was actually a good teacher, although I do not think that he took a great interest in teaching, and his position at the Carlsberg Laboratory did not give him much opportunity to teach. Winge was undoubtedly also a bit of a tyrant. However, he lacked one more qualification for being the head of a school. Despite his lifelong mycological collaboration with his friend Ferdinandsen and despite his close collaboration with Catherine Roberts during the later years of his yeast work, Winge was fundamentally a »lone wolf«. He was actually a rather shy man, an introvert, to whom oral communication with others probably never came easily, and actual team work was alien to his personality. When facing a large audience Winge had the manners of a rather reserved aristocrat, and this was undoubtedly also the impression of casual visitors.

I was fortunate enough to work in Winge's laboratory as a student from 1933 to 1936 when the yeast work was in its most exciting period. My best memories of Winge are from the evenings when he would drop into my lab, sometimes smoking a long pipe (he always claimed that this was the only type of pipe which one could smoke when doing microscopy), and

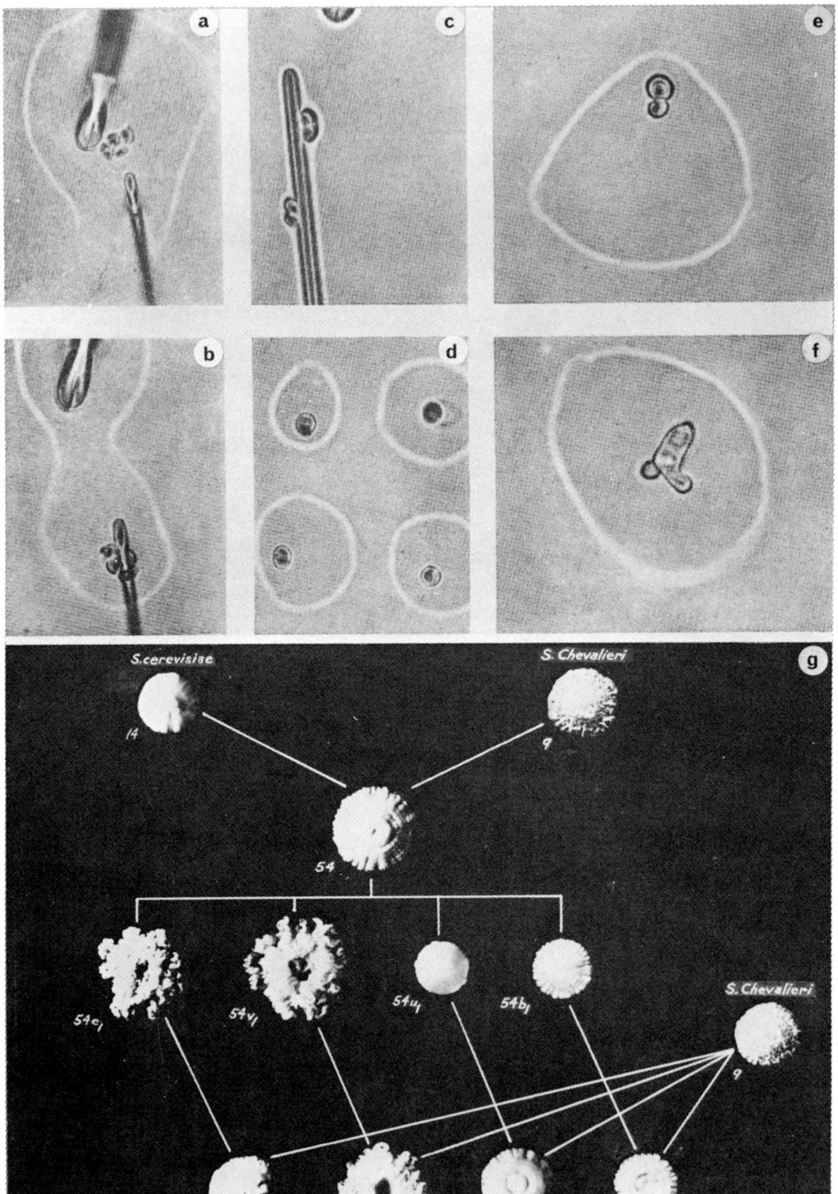

58 Hybridization of yeast.

a–d: Micromanipulator technique for isolation of all four spores from an ascus of Saccharomyces. The spores are isolated by means of two glass needles, one about 7 μm, the other about 2½ μm thick at the point.

e-f: A spore of Saccharomyces validus (left) and a spore of Bakers' yeast always accompanied by his dog »Claus« (who immediately began to chew my shoelaces). He would often sit on my desk, talking for hours, completely relaxed, witty, ironical, discussing genetics and geneticists, a goldmine of information – and very, very helpful.

Winge's reputation among his Danish colleagues was very high and, especially in his younger years, ample use was made of his administrative abilities. He served on a number of important government committees and for years he was an energetic president of the Danish Mycological Society and later of the Danish Botanical Society; both societies later acknowledged his services by electing him an honorary member. Winge was elected a member of the Royal Danish Academy of Sciences and Letters in 1927, and from 1946 until he retired in 1962 because of failing health, he was chairman of its section of Natural Sciences. Perhaps the greatest administrative service which he rendered to Danish Science was to serve as chairman of the editoral committee which was responsible for the publication of Jakob E. Lange's »Flora Agaricina Danica«. Without Winge's enthusiasm and prestige, this costly and monumental work – which actually turned out at the end to be a financial asset – would hardly have been published.

After the death of his close friends Ostenfeld (in 1931) and Ferdinandsen (in 1944), Winge had few social contacts with Danish biologists. His recreations were painting when he was younger, and later hunting and bridge, and he found most of his friends among his fellow huntsmen. It was typical of Winge that he could not help mixing his science with his recreations. He was very interested in the breeding of hunting dogs, and he wrote a comprehensive monograph on the inheritance of

(about 3 μm diameter) placed together in a small drop of culture liquid (e), and producing a bud, after conjugation (f).

g: Giant colonies of hybrids between Saccharomyces cerevisiae and Saccharomyces Chevalieri. From the hybrid (54) are derived single four spore cultures which have been back-crossed with S. Chevalieri (diameter of largest culture about 1.5 cm).

coat colour in dogs. This book was later translated into English by Dr. Roberts.

Winge was one of the few Danish biologists of his generation whose work gained world-wide recognition. His international reputation is reflected in the long list of scientific honours which were bestowed upon him. This list includes the Foreign memberships both of the Royal Society (London) and of the National Academy of Science in Washington.

When Winge travelled abroad on lecture tours or to congresses, often accompanied by his charming wife, Mrs. Julie Winge, he could again be somewhat formal and unapproachable when facing a large audience. However, in a small circle it was obvious that Winge was among colleagues who not only admired and respected his work, but who were also his personal friends, and he, on his part, both appreciated and treasured their friendship.

In 1952 Winge had to undergo surgery for a stomach ulcer and after that operation he began to feel his age. During his last years he was ailing physically, but not mentally. He died in Copenhagen of a lung embolism on April 5th, 1964. He is survived by his widow, Julie Winge née Begtrup Møller and by their son, Per Winge, a Major in the Danish Army. A daughter was killed in an accident many years ago.

The monument of Winge's lifework is more than 135 publications on mycology and cytogenetics. Winge did not belong to the type of scientists who, as a matter of routine, put their name on their student's publications. Each publication which has Winge's name on it is, in every sense of the word, his own work. Some of his investigations, e.g. the theory of polyploid evolution through hybridization and chromosome doubling, his work on sex-linked inheritance and sex determination in *Melandrium* and *Lebistes,* his pioneer work on chromosome patho-cytology and the papers on yeast reproduction and genetics, will always be listed among the fundamental papers in genetics.

Niels Nielsen

1900–1960

by *Poul Larsen*

Niels Sophus Lorentz Benzon Nielsen was born on December 22, 1900 in Copenhagen as the son of Niels Nielsen, Professor of mathematics at the University of Copenhagen, and his wife Ebba, née Benzon-Hansen. He finished high school in 1919 and received the degree of mag. scient. in Botany at the University of Copenhagen in 1925.

As a graduate student he started investigations in the field which was to be the subject of his doctoral thesis, namely substances regulating cell elongation in plants, later to be termed auxins. Nielsen began these studies in the Laboratory of Plant Physiology, University of Copenhagen, with P. Boysen Jensen, and he published a paper on this subject in 1924, and one with Boysen Jensen in 1925. After graduating, he was appointed scientific assistant at the Laboratory of Plant Physiology of the Veterinary and Agricultural University in Copenhagen, with Frederik Weis. In the meantime he had become interested in fungi, partly as pathogens. This led to a fortunate combination of the study of the physiology of fungi and of seed plants, when Nielsen began to investigate the effect of certain undefined metabolic products of fungi on the elongation of oat coleoptiles. He found that pieces of the agar substrate on which the mycelia of certain fungi, namely Rhizopus sp. and Absidia ramosa, had been growing, would produce curvatures in decapitated oat coleoptiles when applied unilaterally to the cut surface. The curvatures were directed away from the agar block, so the material therein must be stimulatory to elongation. The implications of this discovery were elaborated in investigations published in Nielsen's doctoral thesis (1930). He cultivated Rhizopus suinus on a larger scale, permitting an extensive study of the active material, which he called rhizopin. He purified the rhizopin to a considerable extent but without achieving a chemical identification. In view of later developments in the field, there is no doubt that rhizopin was indole-3-acetic acid, subsequently established as a true auxin, a natural regulator of cell elongation in the higher plants. There is thus a striking parallelism between the early research in auxin chemistry and in the studies on the gibberellins. Both were obtained from fungi before they were identified as the native growth regulators in higher plants. Much later, in 1958 and 1959, Nielsen in fact took up the study of some effects of gibberellic acid on seed metabolism.

In 1926, Nielsen spent a period of study in Würzburg, and in 1929 he was appointed head of the research laboratory of the Carlsberg Breweries. In 1931 he returned to basic research, when he was appointed head of the newly established Sub-Department of Fermentation Physiology.

In accordance with the purpose of his Sub-Department, he directed his research activities mainly toward growth substances promoting the increase in dry weight of microorganisms. His main associate was Vagn Hartelius, co-author of numerous publications with Nielsen. In addition, a number of visitors from several foreign countries and Danish institutions worked in the department and contributed to the long list of its publications.

Already in 1932 Nielsen and Hartelius proposed a nomenclature of growth substances which became widely adopted. The rhizopin-like substances which stimulate cell elongation in coleoptiles and stems, but retard or inhibit the elongation of main roots, were called *Group A,* thus alluding to the name *auxin,* which in the same year was given by F. Kögl to this group of growth substances. Substances which stimulate the increase in dry weight during the growth of microorganisms, on the other hand, were called *Group B.* This group of sub-

59 Niels Nielsen, 1930.

stances resembles the *bios* of E. Wildiers (1901). In addition to rhizopin, Nielsen had demonstrated the production of Group B substances in Rhizopus suinus. They stimulated the growth (dry matter production) of Aspergillus niger. Group B substances were present in several other organisms and materials, e.g. beer wort.

Later Nielsen and Hartelius (1937) found it necessary to

distinguish two sub-groups of growth substances under the main Group B, namely *Group B1*, substances which are easily destroyed by oxidation and stimulate the dry matter production in yeast, and *Group B2*, substances which are stable to oxidation and stimulate the dry matter production in Aspergillus niger and other moulds. Substances in Group B2 require the presence of a number of metallic ions as co-factors for their activity. Most of the bios-substances have subsequently been identified as vitamins, vitamin constituents, metabolites, amino-acids, or co-factors for specific enzymes functioning in cell metabolism.

Much of Nielsen's subsequent work dealt with substances in the bios group. He developed a method for the separation of growth substances of Groups B1 and B2. Yeast will not grow without an external supply of some B1 substance, normally taken up from beer wort. The yeast will absorb its indispensable B1 factor so vigorously that a sample of wort after being shaken for 3 hours with a comparatively large quantity of yeast, would not support the growth of a fresh inoculum of yeast. The treated wort, however, was still able to stimulate the growth of Aspergillus niger, and thus contained one or more Group B2 substances.

These studies were extended by Nielsen and Fang Sing-Fang from Tientsin, China (1937). They worked with 3 species of mould and 6 species of yeast. Crude extracts of the moulds, of the mushroom Boletus edulis, and of yeast would stimulate the growth of yeast and, to a smaller extent, of the moulds. Wort shaken with baker's yeast had only a slight stimulating effect on most of the yeasts, but a marked effect on Aspergillus niger (almost as strong as that of unshaken wort) and some effect on Penicillium and Rhizopus. After heating a solution of glucose and ammonium tartrate, the mixture stimulated the growth of Aspergillus niger, but not that of the two other moulds or of yeast. The heated mixture thus acted as a Group B2 factor. In 1940, Nielsen and Jonas Dagys from

Kaunas, Lithuania, showed that the two compounds could be replaced by other monosaccharides plus the ammonium salts of other organic acids.

Several amino-acids had been shown to act as growth substances for yeast, and among these, β-alanine, studied by R. J. Williams and E. Rohrmann (1936) and W. L. Miller (1936), seemed to occupy a special position. It was active in extremely low concentrations ($1{:}10^9$, Nielsen and Dagys 1940) and did not function as a source of nitrogen for the yeast (Nielsen 1936). In 1938 Nielsen and Hartelius started the publication of an extensive series of studies on the effect of amino-acids on the growth (dry weight increase) of yeast. In the course of these investigations they found that β-alanine was stimulatory only when thiamin and aspartic or glutamic acid (or a corresponding amide) were also present. A study of the complicated relationships between the effects of these substances and the incorporation of β-alanine into pantothenic acid was later continued by Hartelius (1946). It may be mentioned here, however, that several amino-acids, including β-alanine, when added singly were found to be inhibitory to the growth of yeast.

In cooperation with some of the visitors to the laboratory, Nielsen and Hartelius have developed biological methods for the quantitative determination of various growth factors. β-alanine was first determined with yeast as the test organism (1941), but later (Nielsen, Hartelius and G. Johansen 1945) both pantothenic acid and β-alanine could be determined using Streptobacterium plantarum in a new method, the »compensation method«. Quantitative methods for the determination of biotin were developed using yeast or Bacterium radicicola as the test organism.

These and other methods were used in investigations of the growth factor and vitamin requirements of various microorganisms and their ability to synthesize vitamins or their precursors.

It had been known since 1941 (T. Wieland and E. F. Möller)

that at least one strain of yeast was able to combine β-alanine and pantoic acid to pantothenic acid. The strain of yeast (Saccharomyces cerevisiae C. L. 1) studied most extensively by Nielsen and his coworkers with respect to growth factor requirements behaved similarly. It will grow equally well with β-alanine and with pantothenic acid as the added growth factor. Evidently, the much studied β-alanine is not active as such, but only after it has become incorporated in pantothenic acid; and even the latter has to be combined with other building blocks to the complete coenzyme A before it can perform any useful function, as shown much later in other laboratories.

A similar study of thiamine showed that C. L. 1 required either the whole molecule or just the pyrimidine moiety, whereas thiazole, added alone, had no effect. This yeast strain is evidently able to combine the two moieties.

In the years 1943 to 1945, Nielsen, Gordon Johansen and Knud Roholt carried out investigations on the mode of action of growth factors and vitamins using the principle of antivitamin activity. Several compounds, each differing chemically from the active molecule in some presumably essential feature, were tested. It was found that pantothenic acid analogues acting as antivitamins in tests with Streptobacterium were not the same as those acting on yeast. None of the compounds tested acted as antivitamin in tests with Aspergillus and Torula.

With Professor George Hevesy and, in part, Professor K. Linderstrøm-Lang, Nielsen has studied the exchange between yeast and its culture medium of phosphorus, potassium, iodine, and bromine, using radioisotopes.

Over the years, several other problems were studied in the Sub-Department of Fermentation Physiology: The agglutination of yeast cells, autolysis of yeast, the action of animal sex-hormones on yeast and moulds, the variations in the content of growth substances in seeds and fungus spores with the age of the seed and spore samples, growth substances in the bleeding sap of birch, the content of nitrogen in individual barley grains, the respiration of germinating barley and of moulds, the effect of vitamin K on yeast, and the fixation of nitrogen by various strains of bacteria.

In addition to papers reporting primary experimental results, Nielsen has published several review articles on growth substances and the physiology of microorganisms, on nitrogen metabolism in fungi (1958) and on nitrogen heterotrophy (1959). He has attended and given lectures at a large number of meetings, in Denmark and in other European countries. He took part in the first international conference on plant growth substances in Paris, October 1937, where 8 leading scientists in the field met to discuss common problems.

Nielsen, who had developed sympathy for the German cause, left the Sub-Department of Fermentation Physiology when the war ended on May 5, 1945. The Sub-Department was terminated as a separate unit on June 1, 1946, when its functions were taken over by the Department of Physiology. In September 1945 Nielsen married the nurse Karen Marie Gudrun Heeris. He moved to Sweden, where he became associated with the pharmaceutical firm AB KABI, Stockholm, from 1949 as laboratory chief. He was lecturer (docent) at the Royal Technical University, Stockholm, from 1948.

In Stockholm he took part in investigations on the chemical constituents of pollen of various wind-pollinating plants which produce large quantities of pollen, particularly with respect to the possible occurrence of vitamins. He also studied the production of vitamins by the yeast Rhodotorula gracilis and took part in investigations on the effect of gibberellic acid on the metabolicm of germinating barley.

Niels Nielsen died in Stockholm on September 6, 1960.

Nielsen and his associates have contributed much information on the general nutrition of microorganisms, in particular yeast. They have made important contributions to our knowledge of the growth substance requirements of several microorganisms and the differential ability of individual strains to synthesize growth factors. Nielsen's laboratory became widely recognized as a center for the study of growth substances of the bios group and will be remembered as such. Nielsen himself will also be remembered for his demonstration of the formation of auxin by the fungus Rhizopus suinus.

Heinz Holter

by *Erik Zeuthen*

Heinz Holter was born on June 5, 1904 in the village of Leonding near Linz in Upper Austria. As a boy he lived with his parents in Graz but he got to know remote places in the Austro-Hungarian double-monarchy during World War I, because the family was on the move with the father, a physician then serving in the army medical corps. Both parents died around the time when Heinz Holter finished grammar school, and he was taken into the house of near relatives of the father in Wels in Upper Austria. Heinz Holter's branch of the family resided in this town for generations as craftsmen, and also Heinz feels that he has his roots there.

Holter studied chemistry at the University of Vienna from 1922 to 1928. His main teacher was the organic chemist Ernst Späth who worked with plant alkaloids. Holter graduated with honours in 1928, and his doctoral thesis was about alkaloids in Corydalis cava (fumitory). Holter became Späth's assistant and continued in the laboratory, but the problems of chemistry did not satisfy him. He had developed an inclination towards biology and began to read about enzymes. He pondered over the function of the enzymes when they were part of cell structure, as distinct from their function in free solution. He coined the term »topochemistry« for his new interest but had to put it aside for the time being. Holter qualified for a Rockefeller fellowship and chose to study in Copenhagen. This led to Heinz Holter's first meeting with S. P. L. Sørensen and Kaj Linderstrøm-Lang which he has described in another chapter of this book.

While modern cell biology is dominated by problems about the relation between structure and function, this was not so when Holter came to the Carlsberg Laboratory, and it was in this field that he and Linderstrøm-Lang put their efforts in the thirties. In this decade they aimed at understanding physiological functions of tissues and of single cells by quantitatively accounting for the distribution of enzymes in cells and types of cells in tissues, and in organels and cytoplasmic matrix in single cells. However, the necessary analytical tools were non-existent and had to be developed so as to permit work on a scale 10^3–10^4 times smaller than that of conventional methods. First they published titrimetric methods for use mostly with tissue sections, then followed more sensitive gasometric methods for determination of enzymes in single cells and parts thereof.

The »Carlsberg Methods« for quantitative work in the μl range that were thus developed consisted of a whole system of microburettes, micropipettes, μl reaction vessels (including the Cartesian diver), devices for filling and handling them, magnetic stirring devices, etc., etc. They were invented through years of continuous, inextricable collaboration between Linderstrøm-Lang, Holter and their co-workers. I think it can safely be said that many of them received their final serviceable form in Holter's hands.

In many laboratories such a set of novel devices would have been patented, but this was against the rules and traditions of the Carlsberg Laboratory. But if they were not a commercial success, the Carlsberg methods certainly were a scientific one, and became the main reason for the flood of visitors that began to invade the laboratory from 1930 onwards.

Histochemistry

In the years 1931–40 Heinz Holter, Kaj Linderstrøm-Lang and co-workers published 33 papers under the common heading »Enzymatic Histochemistry«. The series continued far into the next decade without this general title. It deals with micromethods for the measurement of the enzymatic hydroly-

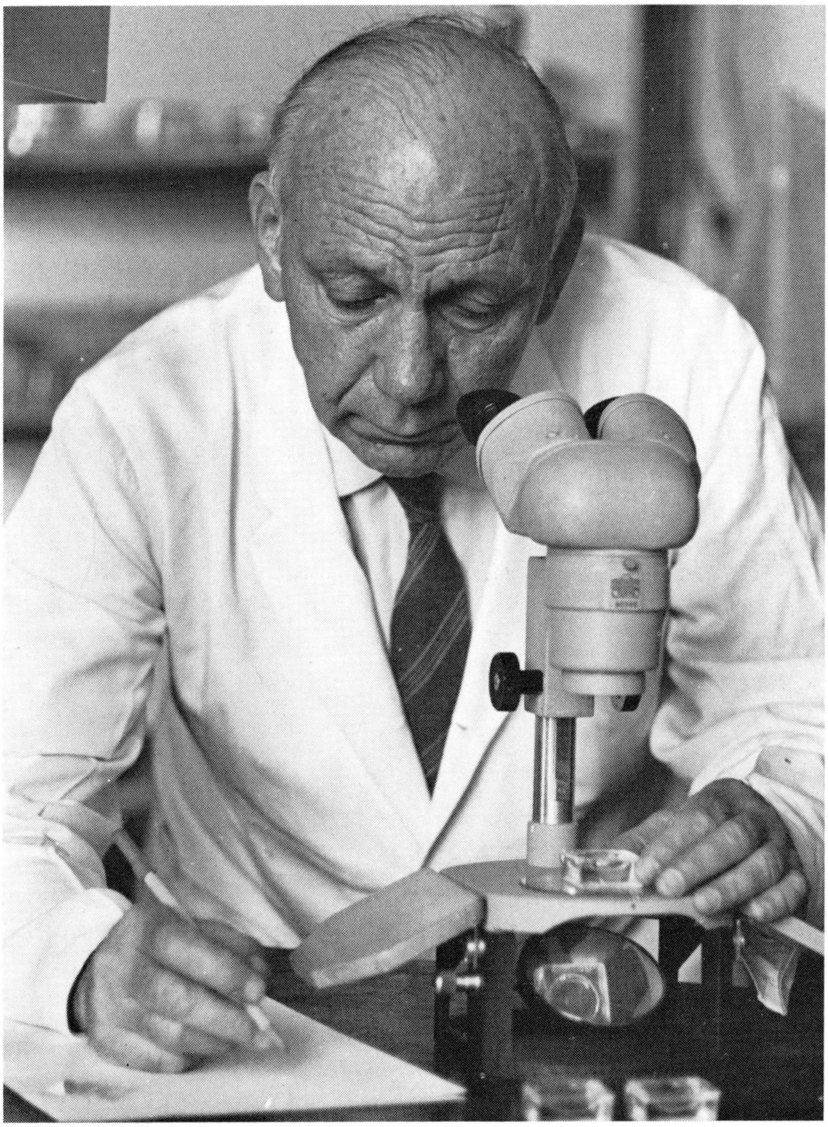

60 H. Holter, 1964.

sis of dipeptides, protein, carbohydrate and fat, and it describes applications of the methods to suitable objects.

The first object to be studied by the new methods was the root of germinating barley, suitable for the purpose because of its size, cylindrical form and differentiation from top to base

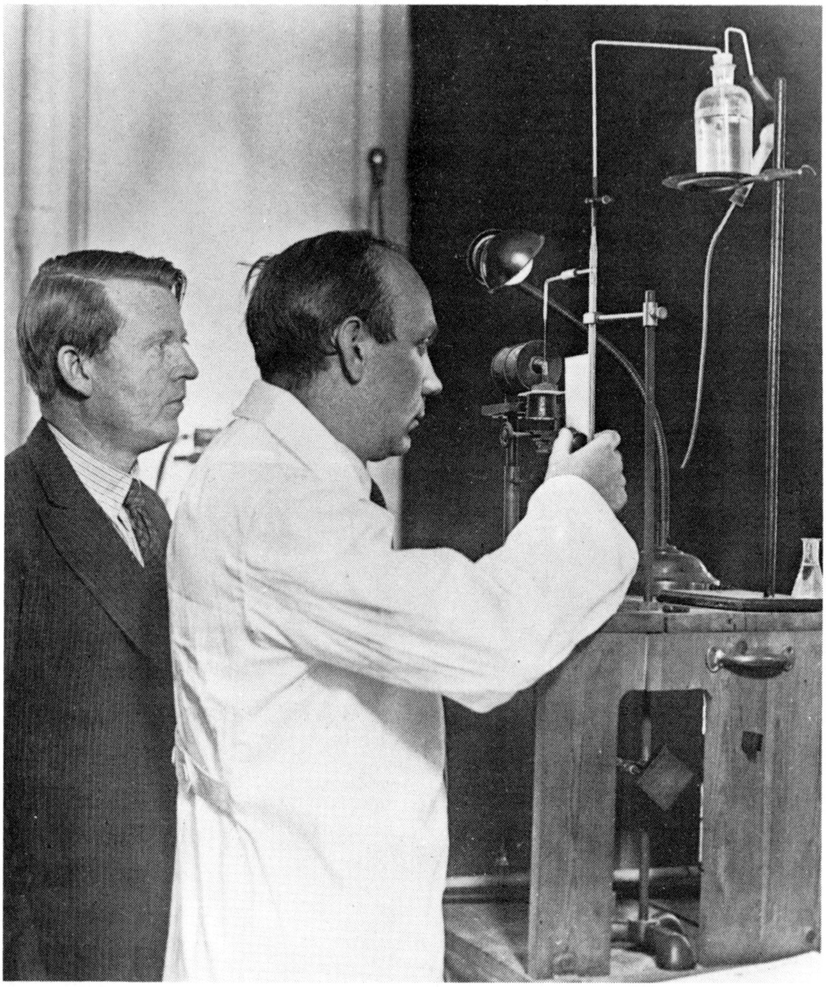

61 Linderstrøm-Lang and Holter during a micro-titration, about 1936.

into zones with differences in cellular activities, and because it can be cut perpendicularly to its axis so as to yield sections of equal thickness for enzyme determinations. The enzymes selected for studies were two dipeptidases previously characterized *in vitro* by Linderstrøm-Lang. At that time the biological function of peptidases was obscure, and Linderstrøm-Lang considered (then and much later) that *in vivo* these enzymes might be engaged in synthesis as well as in hydrolysis, and in germinating barley roots they were indeed found to be confined to the growing parts. Through the years this and related objects were repeatedly studied, and two proteases from the glands of the carnivorous plant Drosera rotundifolia (sundew) were characterized.

62
A »Carlsberg constriction pipette«.

Holter and Linderstrøm-Lang then turned to mammalian tissues, froze them, cut out cylinders and sectioned them while frozen by a technique resembling that used with barley roots. The tissue chosen for study was the stomach mucosa from pig. The purpose was to resolve the problem then current regarding the contributions to the stomach secretion of the three types of cells in the crypts: neck cells, chief cells, parietal cells. There were technical problems at all levels: freezing, precise sectioning, determinations of enzyme and chloride in sections only a few cells thick, and finally problems of cell identification and counting, since chemical measurements destroy cell morphology, and fixation for cell identification destroys the enzymes. This difficulty was overcome by reserving alternate sections for quantitative chemistry and for quantitative histology. Correlations between cell counts and chemical measurements across the mucosa indicated that the parietal cells secrete hydrochloric acid and the chief cells pepsin.

This histochemical mapping of the stomach mucosa paved the way for numerous studies in the field of normal and pathological physiology of the mammalian intestinal tract and of other tissues. Visitors joined such studies at the laboratory, and continued them afterwards. I mention only three, the Danish pathologist A. Søeborg Ohlsen and two guests from the United States, David Glick and Oliver H. Lowry. Lowry's quartz fibre balance and his micro-determination of protein are widely used in many fields of biology to this day. At that time these methods covered an acute need of techniques for measuring sample size. Lowry acquired fame for his further technical contributions and for their application in the topochemistry of brain tissues. Glick became the inventor of a multitude of histo- and cytochemical procedures, the publisher of relevant serials, and the author of the first book on Techniques of Histo- and Cytochemistry (1949). It is proper here to cite his dedication of that book to »Kaj Linderstrøm-Lang and Heinz Holter of the Carlsberg Laboratory, Copenhagen, in appreciation of their scientific achievements, which have contributed generously to the development of histo- and cytochemistry, and in appreciation of their fine human qualities of integrity, understanding, and a truly civilized sense of values and of humor«.

The Carlsberg micro-methods became widely applied also in other fields, for instance during World War II for the studies of penicillin, then available only in minimal amounts. When E. Chain, one member of the team that was awarded the Nobel prize for the penicillin studies, came to Copenhagen after the war, he stated in his lecture: »If the Carlsberg methods had not existed, we would have had to invent them.«

It can be stated quite generally that the availability of these methods induced chemists and biologists to think and to operate in dimensions to which they were not accustomed. The micro-methods helped the biologists to think quantitatively in the dimensions of the cell, and so introduced the concepts quantitative histochemistry and quantitative cytochemistry into biology.

Cytochemistry

The biological problems which brought Holter to Denmark dealt with the enzyme topology of the cell rather than of tissues. The question was: Is the cell a bag of enzymes or are enzymes compartmentalized? – a trivial question to-day but a burning one then. For some time it had been known from E.

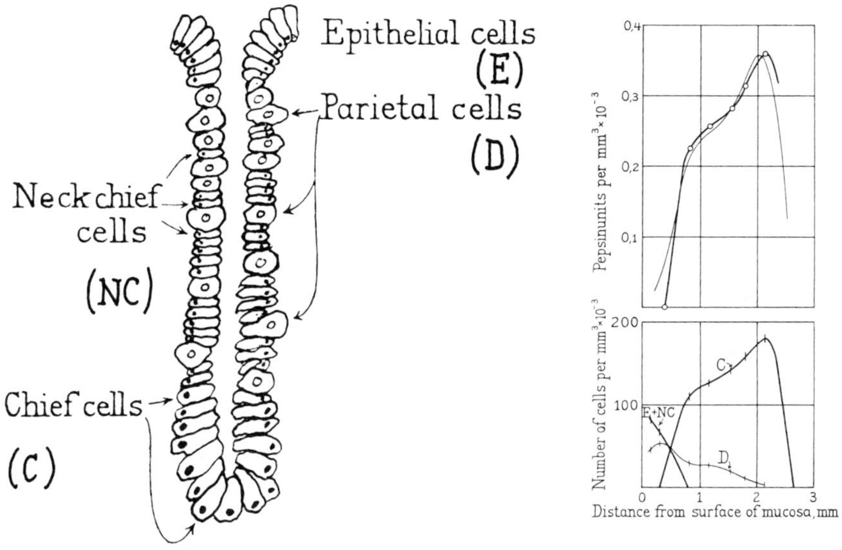

63 The distribution of pepsin in the mucosa of the hog stomach.
Left: Cell distribution in a mucosa gland.
Right: Enzyme concentration correlated with cell counts.

Newton Harvey's work that suitable cells could be stratified alive with respect to granular cytoplasmic inclusions and even torn in two by centrifugal force. When he came to the U.S.A., Holter decided to combine his chemical and Harvey's physical methods in the hope that enzymes would tend to follow specific granular inclusions in centrifuged cells. He went to Princeton and Woods Hole, stratified and partitioned into two Harvey's classical object, the egg of the sea urchin Arbacia, and measured dipeptidase on single halves of the egg. After doing all necessary controls he came to the conclusion that dipeptidase is uniformly distributed in the hyaline ground plasm of this cell and does not move with any of the granular cell inclusions.

If Holter can be disappointed, then it happened to him at this time in the thirties. His chosen marker enzyme did not reveal the secrets which he had hoped for. Discussing the issue more broadly, he and Linderstrøm-Lang could not find conclusive evidence (of their own or of others), that granules known under the name mitochondria were carriers of anything other than the capacity to stain specifically with Janus green and other dyes. Linderstrøm-Lang and Holter did not accept these staining reactions as safe indicators of the presence of respiratory enzymes, and they took no stand relative to the question of whether mitochondria were carriers of enzymes in control of cellular respiration. The Cartesian diver respirometer, which could have supplied the experimental answers, was not yet invented. In retrospect we can see that a great discovery was slipping out of the hands of Holter and Linderstrøm-Lang.

In 1937 Linderstrøm-Lang applied the Cartesian diver principle to micro-gasometry, and through the years of war isolation (1940–45) he and (mostly) Holter developed the method so as to permit manometry of enzyme activities with a sensitivity 10^4 times greater than that obtainable by conventional Warburg manometry. The stage was set for an Enzymatic Cytochemistry based on manometry, and this certainly was to include the respiration enzymes. However, when the war came to an end, Holter found his old problem solved elsewhere. A new era had begun. R. R. Bensley, A. Claude, G. E. Palade and others had homogenized tissue cells, had by the principle of differential centrifugation isolated and identified granular inclusions from the cells, and found that respiratory enzymes do indeed reside in the mitochondria. It seems tantalizing that the very enzyme first examined for specific location in the cell, dipeptidase, to this day has no acknowledged function and no acknowledged site. Dipeptidase is known for what Holter made it, a »ubiquitous matrix enzyme« with unknown function.

Holter's work from the thirties gave him the Cytochemical Subdepartment in 1943. The affix »Sub« is appropriate, because he stayed in the three basement rooms he used to occupy in the Chemistry Department. However, from 1946 he got his own modest quarters in the house Gl. Carlsbergvej 8, where he had his perhaps happiest years. Within the framework of the Chemical Department and its facilities, and preserving close personal and scientific ties with Linderstrøm-Lang, he was now fully on his own. Linderstrøm-Lang was drifting back into protein chemistry, and Holter devoted himself to studies with the giant amoeba Chaos chaos, well named for its bewildering contents of a large range of inclusions in the steadily flowing cytoplasm. Holter hoped that using this cell for *in vivo* separation of cell organelles by centrifugation followed by micrurgical separation of cell halves or even smaller fragments he would obtain fundamental new knowledge, not only about the location of enzymes in the cell, but also about the non-enzymatic chemistry of cell organelles and of integrated cellular functions.

The discovery in the U.S.A. in the early forties of techniques for the isolation of subcellular particles en masse by differential centrifugation of tissue homogenates was of highest importance for Holter, but war conditions prevented him from taking part in this development. After the war he decided to follow his previous line, essentially restricting himself to the study of one well chosen cell, the giant amoeba. In this way he was able to contribute with his small laboratory to the mounting flood of contemporary cell science but at the same time remained sufficiently outside the main current to preserve a chance of being heard internationally. Assisted by Nils Andresen, Holter set out to observe and describe the inclusion bodies and their behaviour in the giant amoeba. Quantitative aspects were emphasized, and the study of normal cells was followed by studies of cells centrifuged alive upon a cushion of high density solution. Under these conditions the cell elongates, the cytoplasm arranges itself in a density gradient, and becomes locally enriched with cytoplasmic granules. The centrifuged amoeba can be cut in two with a glass needle at any place along its axis. The two fragments can then be analyzed for contents of specific enzymes and specific inclusion bodies, and deductions can be made which relate to associations of enzymes with the components of the cell.

Holter, S. Løvtrup and their collaborators found that dipeptidase was uniformly distributed between light and heavy amoeba-halves, while proteinase and succinodehydrogenase were both concentrated in the heavy half. Thus peptidase more or less follows the distribution of the hyaline ground plasm, as concluded from earlier studies with marine eggs, but proteinase and succinodehydrogenase are particle bound. By this time studies elsewhere with mammalian tissues had clearly indicated that succinodehydrogenase – a respiration enzyme – is bound to the mitochondria. To settle this point clearly also with Chaos chaos, Holter and co-workers treated the amoeba the same way as others had treated tissues. They homogenized the cell, concentrated the mitochondria by differential centrifugation and indeed found the enzyme to be contained in the mitochondrial fraction. Thus both *in vivo* and *in vitro* separations of cellular organelles lead to the conclusion that the respiration enzymes reside in the mitochondria. If the Holter approach was not the first to lead us to this important conclusion, it led to its verification; it also proved that the localization in mitochondria is not an artifact of the homogenization technique, and that it applies not only to mammalian cells, but also to phylogenetically far removed organisms like protozoans and the slime mould Physarum.

Cytophysiology

The capture, digestion and assimilation of food, as well as the excretion and defecation of waste products is an exciting chapter in amoeba physiology, and Holter concentrated on this aspect from 1952, working with Nils Andresen, Cicily Chapman-Andresen, Knud Max Møller, John Marshall and others. Chaos chaos was fed radioactively labelled live food, ciliates, which themselves had been fed micro-algae grown in the presence of radioactive 14 C-bicarbonate. The prey is captured in a food vacuole from which water is resorbed and into which hydrolytic digestive enzymes are secreted. The vacuole and its contents are fragmented into smaller vacuoles which after 2–3 days empty non-resorbed contents to the exterior by fusion of the vacuole with the cell membrane. Early users of 14-C autoradiography, Holter and his associates followed the carbon of the food quantitatively through the amoeba for up to 900 hours. Accumulations of 14-C were found in different strata of *in vivo* centrifuged amoebae at different times.

S. O. Mast and W. L. Doyle had observed earlier (1934) that addition of protein to the medium can induce Amoeba proteus to take small volumes of medium into its cytoplasm in the form of vesicles which pinch off from invaginations of the outer membrane. Following W. H. Lewis (1931), they called the phenomenon pinocytosis. Holter and Marshall (1954) wanted to get defined macromolecules into Chaos instead of living prey. They therefore took rabbit gamma-globulin labelled with fluorescein and found that it induced pinocytic vesicles and itself soon appeared in vacuoles, in which it was digested, liberating the fluorescein moiety for excretion. By use of protein (and other inducers of pinocytosis) as vehicles, Holter, Chapman-Andresen and co-workers found it possible to introduce a number of substances that otherwise would not have penetrated into the cytoplasm of amoebae, along with the »drinking fluid«. Together with observations by others on cultured cells, the report of Holter and Marshall constituted a rediscovery of pinocytosis.

In 1956 the life of the Cytochemical Subdepartment came to a close. Øjvind Winge, the head of the Physiological Department at the main laboratory, retired, and Holter was appointed his successor. The department needed a fair amount of technical modernization, and for some time Holter was occupied by administrative tasks. When he was able to return again to cell physiology, Holter for some time turned his attention to cells other than the amoeba, to yeast and mammalian cells.

This choice had to do with the fact that he now felt a certain obligation to work with yeasts. For the yeast work he associated himself with C. Roberts (one of Winge's assistants), P. Ottolenghi, and J. Friis.

In order to utilize his experience with pinocytosis, Holter tried an approach that can now be recognized as rather premature: He and Chapman-Andresen had succeeded in carrying glucose into amoebae impermeable to glucose, using as vehicles pinocytic vesicles induced with gamma-globulin. Could this be repeated with yeast protoplasts and with foreign DNA instead of with glucose? If so, it might become possible to affect genetic transformation of a higher cell. However, experiments (with Ottolenghi, Friis and Kurt Andersen, a biologist from the Carlsberg Brewery) gave negative results. The various types of yeast protoplasts showed no pinocytosis, and other parallel approaches towards transforming yeast were unsuccessful. This early attempt at »genetic engineering« seemed simple and plausible, but it took twenty more years of molecular genetics before genetic transformation of higher cells became actually possible.

Next Holter and Howard Holtzer explored which types of mammalian cells show indications of being capable of imbibing the medium by pinocytosis. Gamma-globulin labelled with fluorescein was the inducer and the marker of pinocytosis. It was concluded that pinocytosis can be induced only in amoeboid types of cells, and this tended to agree with Ch. Pomerat's cinematographic observations according to which amoeboid undulating membrane movements sweep small volumes of medium into the tissue cell.

Thus general acceptance of the fact that molecules to which cells were hitherto considered impermeable can in fact get into cells by membrane flow followed by membrane vesiculation was followed by general agreement that amoebae were useful cellular models for the study of these processes. This made Heinz Holter a busy traveller in reviewing, discussing and disseminating the views developed in the home laboratory. Chapman-Andresen standardized and quantified the pinocytic response in Amoeba proteus and in Chaos chaos, and also tested a large number of chemical agents for their ability to induce pinocytosis. She and others working in Holter's laboratory (Ph. Brandt, D. Dick, K. Hendil) developed the concept that all active inducers of pinocytosis are ionic, and active only in cationic forms. The mucous coat is a cation adsorber or exchanger, and the adsorbed inducer is ingested into the cell in sometimes impressive amounts, when the membrane invaginates.

Holter's studies on phagocytosis, digestion and defecation in amoebae, and his development of the field of pinocytosis led to a new understanding of the dynamics of cellular membrane systems and to developments of the concepts of endocytosis by membrane invagination and exocytosis by fusion of separate membrane systems. From Holter's point of view, the discovery of the lysosomes by C. de Duve (1959) supplied the link which he had so far lacked when taking a broad view of the inter-cellular digestion as he knew it from the amoebae. Digestive vacuoles had long been generally known; Holter had long placed great emphasis on his many observations of coalescence between vacuoles in the cytoplasm of Chaos; however, even though it was supposed that injection of hydrolytic enzymes into food vacuoles would change these into digestive vacuoles, the injection itself had not been observed before. de Duve's discovery once more showed the strength of the technique of tissue homogenization followed by differential centrifugation, and once again Holter and his collaborators quickly complemented the field by cytochemical studies with *in vivo* centrifuged amoebae, demonstrating the presence in Chaos of lysosomes, and presenting evidence that these vesicles fuse with food vacuoles and turn them into digestive vacuoles. Thus essential similarity between pino- and phagocytic digestion in amoebae and autophagic digestion in tissue cells was established.

Membranes ingested by pinocytosis soon change their properties, in the case of Chaos chaos from low to high permeability for glucose. Can they also leak certain macromolecules into the cytoplasmic matrix to serve specific physiological functions? Many would think that this might be the purpose of pinocytosis, but evidence is still scanty as it was in the days when Holter tried to transform yeast protoplasts.

Other work

The above describes what I consider to be the main lines in Holter's work at the Carlsberg Laboratory. There were also many sidelines.

After his arrival at the Carlsberg Laboratory, Holter first began to investigate the peptic digestion of casein. He continued this work until 1935, extending it to comprise the activation of pepsinogen, the kinetics of rennin (chymosin) action, and (with Bent Andersen) a comparison of pepsins and rennins from different mammalian sources. As Holter was working on histochemistry at the same time, he split his fellowship year in New York (1935) between the enzyme chemist J. H. Northrop at the Rockefeller Institute and Robert Chambers at New York University and Woods Hole, between crystalline enzymes and living cells to be operated upon micrurgically.

Holter's work with centrifuged eggs stimulated a lasting interest in embryological problems which led to repeated visits to zoological stations in Roscoff (in Brittany) and Naples. With P. E. Lindahl he followed enzymes and respiratory rates in half-eggs and in the developmental products of separate blastomeres. I myself took part in embryological investigations at the Cytochemical Department and in Naples in the early forties. Holter's assistant, Søren Løvtrup, later took this interest over.

Holter's interest in the amoeba crystals and their formation in, and elimination from the cell continues through his long life with the amoebae. However, it was Holter's co-worker,

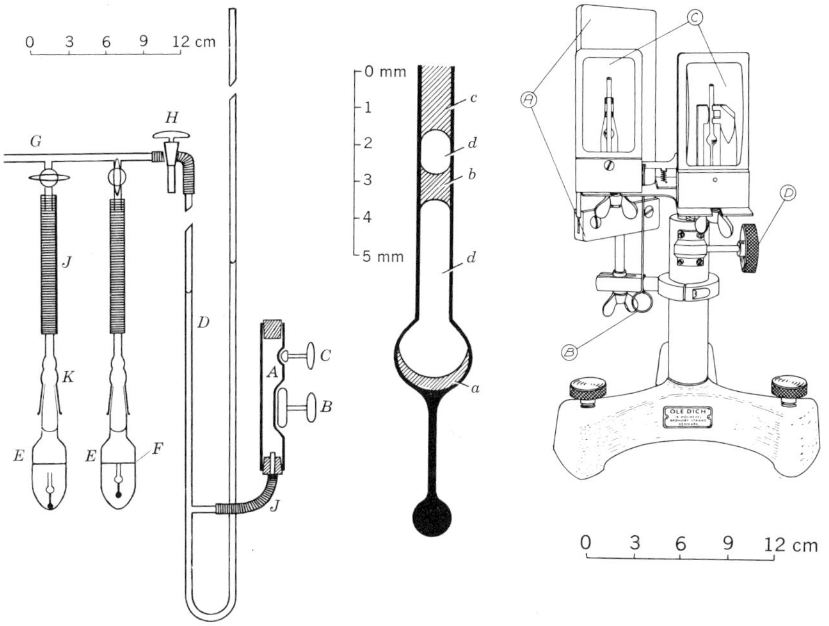

64 Equipment for Cartesian diver measurements. Left: Manometer measuring device (schematical). Two flotation vessels (E) connected through G and H with the water manometer D. The pressure is regulated by means of the coarse and fine screws B and C, compressing the water-filled rubber tube A.

Middle: Standard Cartesian diver (with tail to keep upright), filled with the reaction solution (a) and an oil seal (b). c is the flotation medium extending into the diver neck, d denotes the gas spaces, the volume of which is adjusted in the manometer measurements.

Right: Diver clamp stand (commercial model). A: mirror; B: handle for operating for the diver clamp; C: adjustable magnifying lenses; D: vertical movement.

Knud Max Møller who together with several collaborators produced the foundation of the work which led to their chemical identification as triuret, a major and crystalline product of amoeba nitrogen metabolism. Max Møller and Holter, in close collaboration with the firm »Pharmacia«, introduced Ficoll as a material for producing gradients of low osmotic pressure for the separation of cells and cellular particles by equilibrium centrifugation, a principle which had its origin in earlier work by Holter, M. Ottesen and R. Weber.

Throughout his career Holter approached histo- and cytochemical problems by devising many micro-methods for quantitative analysis. All of these cannot be enumerated here, and only one will be mentioned, the Cartesian diver. After Linderstrøm-Lang's application of the Cartesian diver principle to gasometric measurements, its development turned out to be a very suitable task to carry out during the scientifically quiescent years of the war. When else would it have been done? Moreover, the diver fitted the dimensions of Holter's large amoebae beautifully. In 1944 appeared the final publication about the diver, later popularly called »The Diver Bible«. It consists of two parts, the theory (Linderstrøm-Lang) and the practical method adapted for Cartesian diver respirometry (Holter).

My own contribution at the Carlsberg Laboratory was the Cartesian diver balance. It measures the immersed or »reduced« weight of live cells the size of Chaos chaos and its light and heavy halves. Holter used the diver balance to measure the mass of living cell samples prior to their destruction in analysis for enzymes. A method for measuring another useful parameter, cell volume, was earlier devised by Holter.

The use of the Cartesian diver respirometer spread through many laboratories the world over. Many visitors came to Holter's department in order to learn the use of the diver, and his assistant since 1937, his wife Karen Holter, spent much of her time in teaching these students the use of the diver and other Carlsberg micro-methods.

Heinz Holter, at the age of 26, carried with him to Denmark the diversified multinational culture of Franz Joseph's double-monarchy and of the young Austrian republic. He had lived with arts and music, he had wandered to Madrid to see the Prado, and as a student he had supplemented his meagre resources as a walk-on in the Vienna opera. Although not allowed to raise his voice there, he made music privately, playing the piano and the recorder. He had toured his Austria by bicycle, canoe and on long walks. He knew the life of its peoples, its animals and its plants. He knew it in summer and winter. Coming back from a field trip he may have looked as Linderstrøm-Lang painted him only a few years later (fig. 66). Physically active, reading or always carrying a book with him, he seemed engaged in all aspects of life at the same time.

Coming to Denmark may not have changed Holter's life much. For the Austrian mountains he substituded Denmark's plains of land and water, he became involved with Danish art and music as previously with Austrian, and he found substitutes for Vienna's human multiplicity in the Copenhagener, the Dane, and in that charming cultural mixture to which he himself contributed so much in the Carlsberg Laboratory. From my days of companionship with Holter in Denmark, and not least in and around Naples, I know him to be a perfect traveller with an unusual capacity to observe and enjoy. Whereever he goes, he loves the genuine national traits of which he got so many samples in his first homeland. It is in keeping with this that he himself, when he came to Denmark, made a conscious effort to become a Dane. He was highly successful in this and learned to speak and write our difficult tongue better than most of us natives. He studied and absorbed Danish culture, and became a Danish citizen in 1939. As previously in Austria, Holter hiked at week-ends, and he took with him guests in the laboratory, who in this way got outdoor impressions which complemented those from the hospitable house of Linderstrøm-Lang. These field trips were sometimes strenuous, and occasionally things got out of control, as when Fritz Lipmann one evening in the spring, on the way back from the forests of Sealand, stopped and declared: »Also, Herr Holter, eigentlich bin ich doch ein Stadtmensch«. In summer, week-ends and vacations were spent on cruises with »Susanne«, a sailing boat named after Holter's daughter by his first wife. Susanne her-

65 A meeting of the Danish Natural Sciences Research Council (1969). During Holter's chairmanship the meetings were held in the assembly room of the Carlsberg Laboratory. The persons shown are listed in chapter 14.

self was along, and so was Karen, his second wife, and their son Peter.

Heinz Holter can observe and listen and pass sound judgment. Thus Linderstrøm-Lang consulted him in all matters of departmental significance (and probably many others) over the twenty years when I could observe them, and I have seen Holter used the same way by the Director of the Stazione Zoologica in Naples, Reinhard Dohrn. Holter was elected to the Danish Academy of Sciences and Letters in 1944, has served as its treasurer since 1953, and has been on the board of the International Society of Cell Biology. In Copenhagen he served his term as chairman of the Biochemical and Biological Societies. During the last three years of his professorship at the Carlsberg Laboratory he assumed the heavy load of being chairman of both the Danish Research Council for Natural Sciences and the Joint Commission of Research Councils.

66 Holter painted by K. Linderstrøm-Lang, about 1933.

In the Carlsberg Laboratory he arranged most of the seminars over which Linderstrøm-Lang presided, he gave important seminars himself, and he often led the discussions into profitable channels. For some years (1964–71) the University of Copenhagen used him as a lecturer in »Dynamic Cell Biology«, and for many years he was external examiner in Biochemistry at the Technical University.

Holter's indisputable successes in Danish and world science are inseparable from his qualifications as a human being. Engaged, interesting, warm, full of humour and fun, he made himself loved by everybody. Holter has acquired many friends. By the agency of some of these he has been thanked for his scientific and personal contributions through honorary degrees (Copenhagen, Ghent, Bern), by election to membership of many learned Societies, and otherwise.

To know Heinz Holter properly one must know him on vacation. The adjoining photograph shows him with Karen and Peter heading for the beach in »Susanne«s dinghy, perfectly relaxed, the laboratory chores out of his mind, and ready for new adventures on a summer tour of Danish fjords and islands.

67 Karen Holter, Heinz Holter and Peter Holter off the island of Strynø in southern Denmark (1958).

VI

The men behind the scene (The Board of Trustees)

by *A. Tovborg Jensen*

In the document announcing his endowment which J. C. Jacobsen presented to the president of the Academy of Sciences and Letters on September 25, 1876, the first object (A) of the Carlsberg Foundation was described as »to continue and extend the activities of the Carlsberg chemical and physiological laboratory, founded by me in the year 1875, in conformity with the statutes of that institution.« The first object to be mentioned was thus not the support of external academic activities, which has always been regarded by the public as the sphere of the Carlsberg Foundation. This activity was mentioned next in the document, and was termed object (B).

In 1876 and during the next few years the greater part of the relatively modest income of the Carlsberg Foundation did in fact go to the Carlsberg Laboratory.

In his accompanying letter J. C. Jacobsen asked the Academy to select from its members a »permanent Board of men of scientific insight and ability« to manage the Carlsberg Foundation. This Board was to consist of five members, of whom three should be scientists, constituting the governing body of the Carlsberg Laboratory. The three scientists (Barfoed, Panum and Steenstrup) who were elected to the first Board of Trustees of the Carlsberg Foundation were perhaps not the most eminent of the twenty scientists who were members of the Academy in 1875, but as far as can be judged a hundred years later they were the three with the best prerequisites for making a reality of J. C. Jacobsen's tersely expressed intentions for the laboratory.

Subsequently, when the Carlsberg Foundation became the owner of one of the largest industrial concerns in Denmark, and also of large capital resources, and when the activities under object (B) had increased greatly from their modest beginnings, the scientific trustees were chosen for their ability not only to govern the laboratory, but also to control the brewery or the Foundation.

In addition to the three scientific trustees the Academy of Sciences and Letters was to select »either one or two men from outside the Academy who might be expected on account of their knowledge of brewing, or for other reasons, to be familiar with and interested in the tasks of the laboratory«. They were to participate in the work of the Board of Trustees as adjunct members. During 85 of the last 100 years the Academy has used its right to nominate *two* adjunct members to the Board of Trustees. For two periods of five and ten years respectively only one adjunct member has sat on the Board.

J. C. Jacobsen mentions only *men* in his documents. The three scientific trustees have in fact been men for the whole hundred years, but in 1972 the first woman, Professor Ebba Lund, was elected as an adjunct member.

In the period 1876–1976 18 scientists and 13 adjunct members have sat on the Board. The 16 retired scientific trustees have served for an average of 16 years each, while for the 10 adjunct members the average has been 18 years.

Today it is possible to trace only in broad outline what the trustees have achieved year by year, and how the credit and responsibility has been shared between the five individuals. Someone associated with the Carlsberg Foundation once said that the only history which can be based on the minutes of a corporation is the story of what it has seemed fit to be included in the minutes at the time. But if an attempt is made to bear in mind the personalities of the department heads and trustees of a given period, and also the activities and achievements of the laboratory, as shown in its publications, then the dry pages of the minutes acquire a new life. The impression is of a lively and fruitful co-operation, an impression which probably in the main reflects the true state of affairs.

The first chairman of trustees, C. T. Barfoed (chairman 1876–1889) was aged 61 when nominated. He was professor of chemistry at the Agricultural and Veterinary Univeristy in Copenhagen (founded 1858) and had played a leading part in its planning. He was an agricultural chemist of the school of Justus von Liebig, and an excellent analyst. He had secured a predominant position for chemistry in the curriculum of the new college, the establishment of which had been strongly supported by J. C. Jacobsen as a Member of Parliament. It is also due to him that the chemical laboratories at the Agricultural University were for many years the largest and most modern in Copenhagen.

The second trustee was the physiologist P. L. Panum (trustee 1876–1885). He was 56 when nominated, and died only eight years later. Panum and Barfoed complemented one another admirably. Panum had been professor of physiology at the University of Kiel (at that time a Danish university) for eleven years, but since the war of 1864 between Denmark and Prussia he had been working at Copenhagen University. In both places he had established laboratories for medical and physiological chemistry. For many years Panum's laboratory was the only place in Copenhagen where medical research could be practised. Recently one of the large scientific institutes of the University of Copenhagen was named after him, an honour shared only with H. C. Ørsted, August Krogh and Niels Bohr.

Like Barfoed, Panum never hesitated to undertake outside work in the service of his country. He was concerned with human nutrition and with problems of diet in hospitals and prisons, as well as with pure physiology and microbiology.

The third trustee was the 63 year old Japetus Steenstrup (trustee 1876–1897), professor at the University of Copenhagen. Steenstrup was distinguished by his contemporaries as a »naturalist«, and was applauded to an unprecedented extent as a zoologist, geologist and archeologist. However, this opinion was not shared by a number of people, including the great zoologist William Sørensen, a polemicist and lover of justice: they felt that Steenstrup was not worthy of so much praise, either as a research worker or in his character. Steenstrup established the zoological museum at the University of Copenhagen, and was characterised as »quick at wire-pulling, slow at publication«.

The Academy can hardly have had much choice in nominating the first three scientific trustees. Barfoed and Steenstrup had helped J. C. Jacobsen in drawing up the statutes of the Carlsberg Foundation, and were therefore obvious candidates. Furthermore, no one could be better suited to serve as adjunct members than J. C. Jacobsen himself and his closest associate, the brewer and director Erhard Kogsbølle. J. C. Jacobsen served for twelve years until his death in 1887, and Kogsbølle for thirty years until 1906.

Of the first team of trustees Jacobsen and Barfoed were certainly the most active in the affairs of the laboratory, and have never been equalled subsequently in this respect. It appears from Emil Chr. Hansen's biography that Jacobsen's interest in the laboratory occasionally went so far as to become a nuisance. On the other hand, on his trips abroad he brought back expensive instruments for the laboratory, and he repeatedly met special items of laboratory expenditure out of his own pocket.

The board rapidly established a framework for its activities which has continued without major changes for a hundred years. It worked out the annual budget for submission to the directors of the Carlsberg Foundation, who would then scrutinize it before it was finally approved. The board laid down

C. T. Barfoed

P. L. Panum

J. Steenstrup

E. Warming

68 Members of the early boards of trustees; paintings in the assembly room of the Carlsberg Laboratory.

rules for how accounts and inventories should be kept, and saw to it that these rules were adhered to. It made rules for the purchase of books and periodicals, and each year reviewed scrupulously the inventories and equipment of the laboratory. In 1877 it was minuted, with a long explanatory statement, »that it is desirable that records of laboratory work should be kept and should remain the property of the laboratory«.

The board of trustees met up to four times a year, discussed research programmes annually with department heads, increased salaries at intervals, and was by no means niggardly in making travel grants to heads of departments and assistants. One point was of such vital concern to Jacobsen that he often emphasized it: before beginning work in the laboratory every assistant had to work for three months in all the departments of the brewery »so as to become familiar with all the operations of brewing and to gain practical experience on which the scientific investigations in the laboratory should depend«. On the death of his father Jacobsen had regretfully to give up any idea of an academic education, but his belief in the universal applicability of such an education clearly had its limits!

The most important step taken by the first board of trustees was to found the laboratory house journal, best known by its French name »Comptes Rendus des Travaux du Laboratoire Carlsberg«. Its first issue appeared in 1878, and 46 volumes have been published to date.

The annual and very lengthy budget meeting, to which department heads are invited, has been within living memory the most important administrative event of the year for the laboratory. At this meeting all the threads are gathered together. Scientific plans, questions of personnel, maintenance and modernisation of buildings, purchase of special equipment, are all discussed thoroughly and without prejudice.

The initial staff of the laboratory in 1876 consisted only of two department heads and one laboratory steward. The budget for the following year was 11,400 kroner, and the appointment of two scientific assistants was envisaged. At the time of Jacobsen's death the heads of departments had three scientific assistants and one laboratory steward, and the budget had increased to 22,940 kroner.

In J. C. Jacobsen's time the trustees witnessed two scientific achievements of the laboratory which exceeded anything which might reasonably have been expected from such a modest concern. The first was Kjeldahl's method of nitrogen estimation (1883), and the second was Hansen's cultivation of pure strains of yeast, and their use in brewing, in the same year. The trustees fully appreciated the achievements of both department heads, perhaps a trifle more in the case of Hansen. His fame spread throughout Europe, and his production of substantial journal publications verged on the enormous. Kjeldahl's production was modest compared with Hansen's, hardly more than one tenth measured in numbers of pages, but his papers are so lucid that they can be read with pleasure to this day. His modesty prevented him from publishing most of his investigations on the carbohydrate and protein content of beer: they would have been of great and lasting value.

It is no wonder that Jacobsen, shortly before his death, made plans for building a spacious modern laboratory to replace the original small rooms, but it was left to the next generation of trustees to realize these plans.

At the time of Jacobsen's death Panum had already been succeeded by the chemist S. M. Jørgensen. Jacobsen was replaced by the brewer van der Aa Kühle, and in the following year the botanist Eugenius Warming succeeded Barfoed. Steenstrup continued until his death in 1897 at the age of 84, and was succeeded by the physicist C. Christiansen.

S. M. Jørgensen (trustee 1885–1914, chairman from 1889) was a professor at the University of Copenhagen, and is one of the most eminent Danish chemists. His contributions to the chemistry of complex metal compounds bear comparison with those of Alfred Werner. His school continues to flourish in

Denmark to this day, and has made a unique contribution to the field of complex chemistry. Jørgensen was a conscientious and experienced chairman. In 1901 and 1912 he carried out a codification of the resolutions of the board which for many years has served as a guide to later trustees.

S. A. van der Aa Kühle (adjunct member 1887–1906) had been nominated by Jacobsen to succeed him as managing director of the brewery. His work as director was of decisive importance for the brewery, but he was also very active in the board of trustees for the laboratory, in contrast to his colleague Kogsbølle.

Professor Eugenius Warming (trustee 1889–1921, chairman 1914–21) held a chair at the University of Copenhagen and was a botanist and an ecologist. For a number of years he was the only biologist on the board, and he was very active in getting Johs. Schmidt appointed to succeed Emil Chr. Hansen. His chairmanship fell in the difficult period during and after the first World War, which weighed heavily on him at the age of 72 to 80.

C. Christiansen (trustee 1897–1917) was a professor at the University of Copenhagen and is the only physicist to have sat on the board. He is probably the most eminent Danish physicist of the period between H. C. Ørsted and Niels Bohr. He was unfamiliar with biotechnical industry, but filled his place on the board in virtue of his ability to express essentials in a few simple words, and, as an experimentalist, to get to the heart of a problem with simple resources.

The plans for the new Carlsberg Laboratory could not be realized at once. A building committee was appointed in 1892, five years after Jacobsen's death, and the laboratory was inaugurated in 1897. During this period and in the following years its economic position was favourable, and its capital could be added to each year, in spite of the fact that the new building implied a large increase in running expenses.

Even more important than the new building was the ap-

S. M. Jørgensen

V. Henriques

N. J. Bjerrum

69 Members of later boards of trustees; paintings in the staircase of the Carlsberg Laboratory.

pointment in 1901 of S. P. L. Sørensen to succeed Kjeldahl as head of the chemical department. Sørensen had twice been awarded gold medals by the university for prize essays, and had to his credit a brilliant doctorate thesis. But he was purely an inorganic chemist, and it was undoubtedly a daring move to make him director of a biochemical-technical laboratory. However, a daring appointment can rarely have been more richly rewarded. The results of Sørensen's researches on the chemistry of amino acids, physical chemistry and protein chemistry have acquired a permanent place in the history of chemistry. It is almost incredible that he also assisted the Carlsberg Breweries with practical day-to-day problems, did much work for the rest of the Danish fermentation industry, and even found time to solve problems of national importance for successive governments. The trustees' confidence in Sørensen soon became almost unlimited. For many years they held only one meeting annually, knowing that the Carlsberg Laboratory was in Sørensen's capable hands.

In 1906 Kühle was succeeded by the equally able and enthusiastic Carl Jacobsen (adjunct member 1906–1914), who was managing director of both the Carlsberg Breweries.

In the same year Kogsbølle was succeeded by R. Koefoed (adjunct member 1906–1924), a brewer who had for a short time been co-director of Gamle Carlsberg (the old Carlsberg Brewery). Koefoed was closely associated with the Carlsberg Laboratory, and in fact did scientific work with S. P. L. Sørensen.

The aging Emil Chr. Hansen died nine years after Kjeldahl, and was succeeded by Johs. Schmidt, nominated by Warming. This change caused the trustees a good deal of worry. The physiological department now had to be renewed, but it was not possible to abandon the good work done when Hansen was in his prime, when he had made the Carlsberg Laboratory the Mecca of »zymotechnologists«. The laboratory therefore acquired a supernumerary department head, Albert Klöcker

(1911–1924), who had been Hansen's assistant for many years. The chemical department already had one supernumerary department head, namely H. Schjerning (1904–1914). This reduplication in the laboratory made life difficult both for the trustees and for the established department heads. An additional factor was that Schmidt's research in the traditional fields of the physiological department of the Carlsberg Laboratory was hardly as dear to him as oceanography: he was appointed on the understanding that he could continue with the latter field of research. His contributions to the statutory topics of the laboratory never approached those of Emil Chr. Hansen, let alone S. P. L. Sørensen. However, there can be no doubt about the quality of his work in oceanography. Paradoxically, no investigations have contributed more to impress the name of the Carlsberg Foundation upon the public than Schmidt's »Dana« expeditions and his work on eels. In international oceanography the »Carlsberg Ridge« and the imposing series of »Dana Reports« will have a lasting place.

After the death of S. M. Jørgensen he was succeeded by Hector F. E. Jungersen (trustee 1914–1917), a learned zoologist, who was sceptical of all change. He did not contribute much to the laboratory during his short time on the board. As an oceanographer for many years he presumably looked favourably on Schmidt's activities in this field.

C. Christiansen was succeeded by the great mathematician Johs. Hjelmslev (trustee 1917–1950), a professor at the University of Copenhagen. He was associated more with the Carlsberg Foundation than with the Laboratory.

However in the following year the board acquired in Valdemar Henriques (trustee 1918–1936, chairman from 1921) a man whose business ability and high intellectual attainments made him admirably suited to control both sections. He was a physiologist of the modern school, and had for eighteen years been professor of animal physiology at the Agricultural University and head of the department of animal physiology at the

National Research Institute of Animal Husbandry. In 1911 he succeeded the great physiologist Chr. Bohr as professor of human physiology at the University of Copenhagen.

Warming's successor was another botanist, C. Hansen Ostenfeld (trustee 1921–31), professor at the University of Copenhagen. He accompanied Johs. Schmidt on several of his expeditions. Unfortunately he was seriously hampered by illness during his last four years.

During the first world war the trustees had to battle with great economic difficulties. Salary increases (cost-of-living bonuses) had to be granted at frequent intervals to all the staff, by now quite numerous. The budgetary deficit of the laboratory continued to increase. The government was faced with the same problems, and was almost equally unprepared. Merely to obtain fuel for heating became a matter of life and death. However, the connection between the laboratory and the brewery was an advantage, and the fact that difficulties were overcome owed much to the practical initiative of Koefoed.

Meanwhile, as the war raged, the chemical department flourished under S. P. L. Sørensen. After the war his brilliant young assistant, K. Linderstrøm-Lang, was able to develop his exceptional talents. His achievements also now have a permanent place in the history of chemistry.

As already mentioned, the position was not so satisfactory in the physiological department, especially in relation to Jacobsen's original programme. In 1931 the trustees therefore established a sub-department of fermentation physiology under Niels Nielsen, and acquired for the purpose the adjacent villa Gl. Carlsbergvej 8. This step was not unconnected with the fact that in the previous year the botanist Ostenfeld had been succeeded by Niels Bjerrum (trustee 1931–1956, chairman from 1937), professor of chemistry at the Agricultural University. Bjerrum was one of the most eminent chemists of

70 Knud Jessen, (1938).

his time, and further was particularly expert in some fields in which Sørensen and Linderstrøm-Lang worked. It is sufficient to recall his early electrometric measurements, and his epoch-making work on the acid-base behaviour of amino acids. Bjerrum was therefore unusually suited for the position of chairman. His interest in all the affairs of the laboratory was just as lively as that of J. C. Jacobsen before him. Bjerrum's and S. P. L. Sørensen's work overlapped many times, but the details of their collaboration both in chemistry and in the affairs of the brewery cannot now be disentangled. They had in fact been in close contact before Bjerrum became involved in the management of the Carlsberg Foundation. Linderstrøm-Lang has many times testified to Bjerrum's influence on his work, for example in an article on the occasion of Bjerrum's seventieth birthday.

As a successor to Johs. Schmidt the physiological department gained an authoritative head in Øjvind Winge. His work was central to Jacobsen's original conception, and was scientifically substantial, and at times epoch-making.

The death of Henriques in 1936 was a great loss to the laboratory. He was succeeded by Knud Jessen (trustee 1936–59, chairman from 1956), professor of botany at the University of Copenhagen. Jessen had a reserved personality, but he was very knowledgeable and was equally expert as a field and laboratory botanist of the modern type. The laboratory was of vital concern to him.

For a number of years there had been only one adjunct member, Koefoed. After his death Poul C. Poulsen, technical director of the brewery was appointed (adjunct member 1923–1933) to take care of the relations between the laboratory and the brewery. Shortly afterwards A. C. Andersen, head of the chemistry department of the National Research Institute of Animal Husbandry joined the board (adjunct member 1925–1965). Andersen had been Sørensen's right hand man during Sørensen's first years at the Carlsberg, and was subsequently a very active collaborator of Chr. Bohr. However, after Bohr's early death he had to leave human biochemistry, and rapidly became a leading figure in Danish livestock research.

With Bjerrum as chairman, supported by Knud Jessen, the clear-thinking Hjelmslev and the excellent adjunct members, the composition of the board was almost ideal. Economically they were in calm waters; moreover, Bjerrum's thrift became a legend.

After Poulsen the new technical director of the breweries, N. Hjelte Claussen was elected as adjunct member (1937–1955). Hjelte Claussen was both an industrialist and an eminent fermentation physiologist. He was succeeded by yet another technical director from the breweries, Hans B. Fogh (adjunct member 1955–1964). These last two also strengthened the board to a marked degree.

For many years, under Sørensen, Winge and Linderstrøm-Lang, the light of the Carlsberg Laboratory shone more brightly than ever before. The last two achieved one of the greatest honours open to a scientist in being elected as foreign members of the Royal Society of London. In January 1943 a cytochemical sub-department was established under H. Holter. Cytochemistry was at that time developing rapidly. This department was initially housed with Linderstrøm-Lang in the chemical department of the laboratory, and did not acquire its own laboratories until the closure of the sub-department of fermentation physiology in 1946. The cytochemical sub-department came to an end in 1956 when Holter succeeded Winge as head of the physiological department.

It need hardly be said that the board of trustees was proud of the laboratory. Nevertheless, even those to whom much has been given are never fully satisfied. Winge's research was undoubtedly of the highest quality, but it had less practical significance for the brewery than Emil Chr. Hansen's unique discovery at an earlier date. Enzymatic histochemistry, founded and developed by Linderstrøm-Lang and Holter, solved more problems in animal and human organisms than in barley, yeast or hops. Moreover, the greatness and scope of Linderstrøm-Lang's later work was perhaps not fully appreciated at the time.

The end of the second world war brought economic difficulties for the laboratory in its wake. Equipment became more

71 A. C. Andersen, the adjunct member with the longest time of service of all members of the board of trustees (1925–1965).

72 A. Tovborg Jensen, (1957).

complicated and was very expensive. Salaries, which had been frozen during the German occupation, rose sky high, producing severe problems for the whole of the Danish community. Because of the international reputation of Linderstrøm-Lang and Holter substantial grants for special investigations were received from the Rockefeller Foundation (from 1937) and from the National Institute of Health, USA (from 1960). It should also not be forgotten that the European Recovery Program (known in Denmark as Marshall Aid) made generous grants to the Carlsberg Laboratory in the period 1948–1952 for the purchase of up-to-date equipment. However, the writing on the wall became steadily clearer, and the finances of the laboratory more and more strained.

In 1950 Hjelmslev was succeeded by Børge Jessen (trustee 1950–1963), professor of mathematics at the University of Copenhagen. Jessen was simultaneously chairman of the Board of the Carlsberg Foundation, and was an expert on financial questions; he rendered important services to the laboratory.

The remaining past or present trustees will only be listed here; –

Aksel Tovborg Jensen (trustee 1956–1972, chairman from 1960), professor of chemistry at the Agricultural University; P. Brandt Rehberg (trustee 1961–1970), professor of zoophysiology at the University of Copenhagen; Arne Noe-Nygaard (trustee from 1962), professor of mineralogy at the University of Copenhagen; Christian Crone (trustee from 1971, chairman from 1972), professor of medical physiology at the University of Copenhagen, C. J. Ballhausen (trustee from 1971), professor of chemistry at the University of Copenhagen.

The following have served as adjunct members: –

K. Georg Sørensen (1964–1972), technical director of the Carlsberg Breweries; E. Hoff-Jørgensen (1965–1972), professor of biochemistry at the College of Dentristry, Copenhagen; A. W. Nielsen (from 1972), managing director at the Carlsberg Breweries; Ebba Lund (from 1972), professor of veterinary immunology at the Veterinary and Agricultural University.

The co-operation between the trustees and the department heads during the last 25 years will not be mentioned, let alone described in detail. The economic difficulties of this period are apparent from the following two figures: the expenses of the laboratory were 496,651 kroner in 1950–51, and 2,602,656 kroner in 1970–71. The trustees could rejoice in a number of

73 The Isotope Laboratory, established 1965–66, during the chairmanship of A. Tovborg Jensen.

important scientific achievements from both departments, but this could not budge the hard fact that the economic situation was becoming progressively more untenable. Finally there was a clear choice between curtailing the activities of the laboratory and associating it more closely with the brewery. An account is given elsewhere (Chapter 11) of the deliberations and negotiations which ended in a fortunate solution: the Carlsberg Research Center.

During these years the only item added to the traditional procedure of the trustees was that their chairman, Tovborg Jensen, was required to take part in the periodic meetings of the brewery's research laboratory, which were presided over by Birger Trolle and were also attended by the department heads of the Carlsberg Laboratory.

Although this period was characterized by economy, the trustees were nevertheless able to grant many requests from the department heads for increasing and bringing up to date the technical equipment of the laboratory; each department was also allotted two graduate studentships.

The most costly single item during this period was the construction of a separate building south of the laboratory. This was built at the suggestion of professors Holter and Ottesen, with a special grant from the Carlsberg Foundation, and houses a laboratory for work with radioactive isotopes of low and medium activity.

It is a well known fact that no board of directors can cope with the day-to-day control of a large industrial concern. This can only be achieved by a responsible management led by a man with authority. An analogous relation exists between the board of trustees of the Carlsberg Laboratory and the heads of departments. The trustees can ensure that the laboratory continues to have the right department heads. They can also, to the best of their ability, give advice and assistance to the heads of departments when need arises. But they cannot and should not govern the laboratory.

It can be said with some confidence that during hundred years the trustees of the Carlsberg Laboratory have sought to put the best men in charge of the laboratory, and have supported them to an extent enjoyed by few Danish scientists. In the course of time many Danish university professors have envied their colleagues at the Carlsberg Laboratory from the bottom of their hearts.

However, it is by no means a care-free existence to be head of a department in the Carlsberg Laboratory. Such a man must be a full-time research worker and director of research from his appointment to his retirement, without the possibility of being refreshed by teaching if, in the words of Einar Biilmann »he should pass through a barren period«. Joh. Kjeldahl was weighed down almost to the ground by such a state of affairs. During his last years Linderstrøm-Lang had the notion of arranging a limited amount of teaching at the University of Copenhagen for department heads in the Carlsberg Laboratory. This idea was realised in 1964, when Professor Holter and Professor Ottesen began to give lectures at the university on cytology and protein chemistry. The laboratory has always enjoyed good relations with the university, but regular two-way communication was not established until 1964.

The trustees have had few worries, and these have rarely developed into anxiety. The economic difficulties during and after the first world war, and in the mad sixties and seventies, were far from being the most serious, since they were unavoidable. But in section 12 of the statutes the department heads are required to »train themselves in those scientific developments which are particularly important for the brewing of beer«. Where was the line to be drawn between this requirement and the permission in section 8, «in addition to their work in the special topics of the Institute, they may develop their general scientific proficiency by other studies and investigations«? How much attention should be paid to the warning in the last clause of the paragraph, that »the main aims of the Institute shall not be set aside or lost sight of«? Johs. Schmidt's oceanographic investigations are not the only »studies and investigations« which have given the trustees occasion to reflect on quotations from the laboratory statutes. Some of the trustees followed Gambetta's motto; »Il faut y penser toujours, et n'en parler jamais«.

Another concern of the trustees involved the intention of the founder that every assistant should spend three months in the brewery before starting work in the laboratory. However, this practice was abandoned during the first world war; perhaps it was merely forgotten. It is mentioned in S. M. Jørgensen's 1912 codex, but not subsequently.

The wish of the founder that »the appointment of assistants should aim at training research workers in the appropriate branches of chemistry and physiology« has been amply fulfilled. Many assistants have done substantial and often outstanding work in other institutions after leaving the Carlsberg Laboratory. On the other hand, other assistants have become so firmly attached to the Carlsberg milieu that their dearest wish was never to leave it. Analogous cases occur in scientific institutions all over the world, but they are probably more numerous in Denmark than in most other countries. Danish scientists seem to be particularly reluctant to change their place of work, or even their field of research. This situation has occurred frequently in the Carlsberg Laboratory, probably because its climate of research is particularly fruitful for scientific work, and attractive from a human point of view.

There is no need to emphasize that this problem has troubled the trustees, nor that it has been particularly acute when appointing new heads of departments. On the very few occasions when a difference of opinion between the trustees and the department heads has approached a confrontation it has always involved this kind of problem, for which there is no patent solution. It has also happened that such a problem has not been solved to the satisfaction of all concerned, but it has not always been the department heads or assistants who remained dissatisfied.

Those who have been members of the board of trustees of the Carlsberg Laboratory during the last hundred years have found their work a source of gratification. The trustees have delighted in a series of scientific achievements, which have filled them with pride. They have had the privilige of working with some of the greatest names in Danish science. In the last century, free from economic difficulties, the trustees were able to establish a laboratory which was greatly superior to the Danish standards of that period. Later boards of trustees have witnessed scientific activity and achievements which often far exceeded all reasonable expectations. As a former trustee I can, with joy and confidence, echo the words of the poet: »Happy is he who has seen the old things, and shall see the new«.

74 The present Board of Trustees, (1976).
Left to right: A. W. Nielsen, Chr. Crone, Ebba Lund, A. Noe-Nygaard, C. J. Ballhausen.

Members of the board of trustees

Carlsberg Foundation			*Adjoint Members*	
C. T. Barfoed (Chemistry) 1876–1889	P. L. Panum (Physiology) 1876–1885	J. Steenstrup (Zoology) 1876–1897	J. C. Jacobsen (C. B.) 1876–1887	E. Kogsbølle (C. B.) 1876–1906
E. Warming (Botany) 1889–1921	*S. M. Jørgensen* (Chemistry) 1885–1914	C. Christiansen (Physics) 1897–1917	S. A. van der Aa. Kühle (C. B.) 1887–1906	
C. H. Ostenfeld (Botany) 1921–1931	H. Jungersen (Zoology) 1914–1917	J. Hjelmslev (Mathematics) 1918–1950	Carl Jacobsen (C. B.) 1906–1914	R. Koefoed (C. B.) 1906–1924
N. J. Bjerrum (Chemistry) 1931–1956	*V. Henriques* (Physiology) 1917–1936	B. Jessen (Mathematics) 1950–1963	P. C. Poulsen (C. B.) 1925–1934	A. C. Andersen (Chemistry) 1925–1965
A. Tovborg Jensen (Chemistry) 1956–1972	*K. Jessen* (Botany) 1937–1959	A. Noe-Nygård (Geology) 1963–	N. Hjelte Claussen (C. B.) 1938–1955	
C. J. Ballhausen (Chemistry) 1972–	P. B. Rehberg (Physiology) 1960–1971		H. B. Fogh (C. B.) 1955–1964	E. Hoff-Jørgensen (Biochemistry) 1965–1972
	Christian Crone (Physiology) 1971–		K. G. Sørensen (C. B.) 1964–1972	
			A. W. Nielsen (C. B.) 1972–	Ebba Lund (Microbiology) 1972–

The years indicate the period during which the members served on the board. The names written with italics indicate persons who served as chairmen of the board. C. B. indicates: Director at the Carlsberg Breweries.

VII

Comptes Rendus des Travaux du Laboratoire Carlsberg

by *H. Holter*

One feature of the Carlsberg Laboratory which deserves special mention is its journal, which has played an important role in making the results of the laboratory's work known to the scientific community.

The journal is nearly as old as the laboratory itself. In compliance with the founder's fundamental rule that »no result of the activities of the institute which is of theoretical or practical importance must be kept secret«, the first Board of Trustees (of which J. C. Jacobsen himself was a member) decided as early as 1877 that the results of the investigations carried out in the laboratory should be published in a special journal. It should appear in Danish under the name »Meddelelser fra Carlsberg Laboratoriet« (»Reports from the Carlsberg Laboratory«), but every number should contain a résumé in French. The first number of the new journal appeared in July 1878.

It quite soon emerged (from the fourth number of volume 1) that it was more convenient for foreign subscribers to the journal to let the résumés appear also as separate issues under the name »Comptes Rendus des Travaux du Laboratoire Carlsberg«.

From volume 5 (1901) the résumé issues were discontinued. Instead, each publication was translated in its entirety, so that from 1901 the journal appeared in two separate editions, one in Danish, the other in foreign languages (French, English or German) under the French title.

In this form the journal continued to appear until 1933 (volume 20). The only change that occurred during this period was the gradually increasing preponderance of the use of English in the foreign edition. In 1933 it was decided to discontinue the Danish edition altogether; since then nearly all of the laboratory's publications in the »Comptes Rendus« have been in English.

At the same time another important change was made. Upon the initiative of professor Ø. Winge, newly appointed head of the Physiological Department, it was decided to divide the »Comptes Rendus des Travaux du Laboratoire Carlsberg« into two parallel series, the »Série Chimique« and the »Série Physiologique«. The reason for Winge's proposal was that he regarded the contents of the »Comptes Rendus« in its previous form as too inhomogeneous a mixture of biochemical and physiological communications. He believed that this prevented many people from subscribing to the journal, and that circulation would increase if subscribers had the choice between the two series, each containing papers from only one of the two departments. Accordingly, the two series were issued from 1934, each beginning with volume 21.

In the long run, however, the new measure did not have the success anticipated by Winge. Only few of the subscribers actually chose only one or the other of the two series. Most of them kept both, and for this majority it turned out to be a practical inconvenience that the two series had a rather different frequency of appearance. Further, the purpose of greater homogeneity of contents was only partially realized. Especially in the case of the »Série Chimique«, the interests of the contributors were rather variegated and sometimes papers on undoubtedly physiological subjects appeared in the »wrong« series because the work had been done in the »wrong« department.

For these reasons, after the retirement of professor Winge in 1956, the division of the »Comptes Rendus« into two series was abandoned. At that time, 6 volumes of the »Série Physiologique« had been issued (volume 21-26), while »Série Chimique« had progressed to volume 30. The return to the old order was accomplished by discontinuing. »Série Physiologique« and adopting the sequence of volumes of »Série Chimique« as the sequence for the re-united journal. Beginning with volume 31, therefore, the journal has again been edited under its old title, in only one series. At the present time (January 1976) volume 40 is approaching completion.

In establishing the journal, the trustees realized right from the beginning that it would be impossible to aim at regular frequency of appearance or equal size of individual issues in a journal only sustained by the scientific production of an institution of the size of the Carlsberg Laboratory. It was therefore decided that the issues, while bearing consecutive numbers, should appear at irregular intervals, and that each issue should contain only one or a very few papers. When a certain number of pages (usually in the vicinity of 500) had been issued in this way, the numbers would be collected to a volume. This system is ideally suited to the irregular rhythm of production in a pure research laboratory, and it has been maintained to the present day. Its practical inconvenience lies in the fact that it is impossible to predict whether a volume will be completed during an interval of months or of years, and that subscriptions therefore cannot cover a predetermined period of time, but can only be made by volume.

The circulation of the »Comptes Rendus« has of course always been limited, but not unduly so. Up to the year 1898 each issue was printed in 550 copies, with an additional 100 copies of the résumé alone. After the establishment of 2 editions (Danish and French) the number of copies printed was 750 for each edition. At present, the minimum number of copies printed per issue is 1100; this number may be increased in special cases, when a high demand is anticipated. – A certain number of copies is distributed free of charge to selected libraries and scientific institutions, and a further number used in exchange for journals which are then incorporated in the library of the Carlsberg Laboratory, or given to other Danish scientific libraries.

Although the Board of Trustees in 1877 established the »Comptes Rendus« as the journal destined for the publication of the research carried out in the Carlsberg Laboratory, it has never stipulated that all scientific papers from the laboratory should be published there exclusively. As in so many other respects, the Board has been liberal and understanding in this matter and has not interfered when somebody wished to publish a paper in one of the appropriate international journals, for reasons of wider circulation or speedy communication. But by and large the collected volumes of the »Comptes Rendus« certainly give an adequate picture of the scientific production of the laboratory.

Unlike many comparable »home journals« from other institutions, the »Comptes Rendus« has during most of its long history not published yearly reports covering the general activities of the laboratory. During the first years the Board of Trustees published short reports about the state of the laboratory, not yearly, but at irregular intervals; but this practice has been discontinued since 1923 (volume 14). From then on the Trustees' reports about the laboratory were only published, together with the yearly reports of the heads of the laboratory's departments, in the yearly reports (in Danish) from the Carlsberg Foundation about its activities in support of Danish Science (»Beretninger fra Carlsbergfondet«). By and large therefore, and strictly since 1923, the »Comptes Rendus« has been a purely scientific journal. The only articles of a non-scientific nature that continued to appear in the »Comptes Rendus« were the obituaries of the heads of departments of the laboratory.

Cumulative indexes of authors and contents have appeared in 1928 (covering volumes 1-16) and in 1935 (covering volumes 17-20). A new index will appear at the end of volume 40.

For many years the »Comptes Rendus« has published only work that had wholly or partly been done in one of the departments or sub-departments of the Carlsberg Laboratory, either by the personnel of the laboratory or by visiting scientists. In order to control the quality of the publications, each paper is submitted not by the author directly, but by the head of the appropriate department, who thus assumes responsibility for the scientific quality. Apart from this, there is no referee system. In a few rare cases work from outside has been accepted, but only if it had an especially close relation to the work of the laboratory and was submitted through a head of department.

Since 1956 (volume 31) another institution besides the Carlsberg Laboratory has been included in the circle of contributors to the »Comptes Rendus«. This is the »Biological Institute of the Carlsberg Foundation«, an institution established in 1931 and situated in another part of Copenhagen. The history of this institution is not included in this book, but for the purposes of this chapter it is sufficient to state that the general line of work pursued at the Biological Institute, especially under its present leader, Professor Erik Zeuthen, is sufficiently close to the scientific aims and scope of the Carlsberg Laboratory to warrant its inclusion among contributions to the »Comptes Rendus«.

I have not been able to ascertain whether during the first years of the journal's existence a special editor was appointed to take care of the »Comptes Rendus«. When H. Jessen-Hansen became Kjeldahl's assistant, he seems to have been appointed editor of the journal shortly after 1892, and he continued as editor until his death 1933. After him G. Haugaard took over the editorship until he departed for the U.S.A. in 1939. Since then, and up to the present day, H. Holter has been editor.

This brings the history of the »Comptes Rendus des Travaux du Laboratoire Carlsberg« up to the present date. In conclusion, however, a few words ought to be said about the fate of the »Comptes Rendus« during the recent reorganization of the Carlsberg Laboratory.

Up to 1972, the cost of the »Comptes Rendus« (and it has never been a profitable undertaking) was entirely borne by the Carlsberg Foundation, as a part of the Laboratory budget. When the Breweries took over the laboratory, the »Comptes Rendus« went with it. For some time the fate of the journal seemed uncertain, but finally the new Board of Trustees decided, in cooperation with the Directors of the Breweries, that the journal was worth keeping alive. It was also decided, however, that the time had come for a modernization of the old journal, and the following measures were adopted:

The future name of the journal shall be: »Carlsberg Research Communications«.

The format of the journal shall be somewhat enlarged, in order to give greater scope for variability in size of illustrations.

The circle of contributors shall be enlarged to comprise, besides the Carlsberg Laboratory and the Biological Institute, also the other laboratories of the Breweries' Research Center, provided that the contributions from these other laboratories comply with the rules that so far have governed the contents of the »Comptes Rendus«.

This means, in other words, that the »Comptes Rendus« under its new name and in its new shape will continue with its essential features intact. The »Comptes Rendus« has, despite its restricted contributing clientele, during the 99 years of its existence produced 46 volumes of about 500 pages each.

There is every reason to believe that this rate of output will be maintained or even increased by the »Carlsberg Research Communications«.

VIII

How our visitors saw the laboratory.

The number of visitors to the Carlsberg Laboratory was great, and so was their contribution to the laboratory's achievement. To provide a record of both by means of a chronological list had to be given up as impracticable, even if such a list would have contained interesting names already from early times, such as Louis Pasteur, who visited J. C. Jacobsen in 1884; the discoverer of crossing-over, the Belgian F. A. Janssens, who worked with E. Chr. Hansen in 1893; or the grand old man of Swedish biochemistry E. Hammarsten, who visited S. P. L. Sørensen several times in the first part of this century. But later on the names were just too many, and the idea of a comprehensive visitors' record had to be abandoned. Besides, such a record would scarcely have done justice to the interplay between visitors and staff which was an essential part of life at the laboratory.

It was therefore decided to let our visitors speak for themselves, and the editor has approached a few of our oldest friends who have known the laboratory for many years, and asked them to write about their experiences as they remember them. The very first contacts gave such a positive response that the available space was soon filled up. For this response we are very grateful.

This way of soliciting contributions must naturally be rather selective. It will therefore be found that the period covered is somewhat limited (about 1930 to 1960), and that most of the contributions concern the chemical department. But after all – these were the years and the place which attracted the greatest number of visitors. It may also be found that some of the contributions are tainted too much by friendship and fond memories. In reply to this the editor would like to quote from a letter by the author of one of the contributions: »All exaggerations or speculations are deliberate and you may lay the blame at my door«.

My first meeting with S. P. L. Sørensen

by *H. Holter*

The following contribution to the visitors' chapter consists of recollections from the time when I entered the Carlsberg Laboratory as a visitor, and was originally written as an introduction to a memorial lecture about S. P. L. Sørensen, given in 1968 at a meeting of Scandinavian chemists.

The year was 1929, and I held at that time a position as assistant at the University of Vienna, in the laboratory of the organic chemist Ernst Späth, under whose guidance I did research in pure organic chemistry. My trouble at that time – and perhaps I should not confess it at a meeting of chemists – was that I did not enjoy my work. Without knowing much about it, I was in love with Biology. So when a chance for a Rockefeller fellowship turned up, I thought of going to a biochemical laboratory as an approach to biology.

The Carlsberg Laboratory in Copenhagen was a possible choice but at that time it existed for me only as a vague and

75 The entrance hall of the Carlsberg Laboratory.

foggy notion, until one day one of my friends told me about a small periodical »Comptes Rendus des Travaux du Laboratoire Carlsberg«. So I went to the library and was immediately fascinated by the papers of Sørensen and Linderstrøm-Lang. Within a few days I wrote to Sørensen and asked for his permission to come and learn something about enzymes.

And then, one rainy morning in March 1930 I stood before the Carlsberg Laboratory, somewhat amazed, not to say shocked, by the marmoreal splendour of the building. I was of course quite ignorant of the Carlsberg Breweries, of the Brewer Jacobsen and of the peculiar atmosphere which to this day pervades everything that originated from the activities of that remarkable man.

Furthermore, I did not know that Sørensen's office hours, which stood engraved in gilt letters upon a marble plate, ought not to be taken quite literally. So I had to wait for quite some time till Sørensen actually turned up, and I had plenty of opportunity to get more and more jittery.

Mercifully it is to-day not so easy, at least for the younger generation, to grasp exactly what I was so nervous about. I had been bred and trained in Central Europe, where a professor was an august personage indeed, and to meet such a man for the first time was not a thing to be taken lightly. And therefore it is perhaps also difficult to understand that it was almost a shock for me to discover that here, face to face with Sørensen, I had absolutely nothing to be scared about.

This man was polite, kind, businesslike and without any affectations whatsoever. He had to express himself in German which made for a certain quaintness in some of his statements, but this did not in the least alter the main impression which transformed the meeting to an unforgettable memory: Here was a man marked by so much artless dignity and candour that one could not help trusting him.

There is no reason to recount our conversation. It was short, matter-of-fact, and then speedily terminated by my introduction to Sørensen's assistant, K. Linderstrøm-Lang, who was to be my daily contact and mentor in the wilderness of biochemistry. And thus I entered a world which it is safe to describe as profoundly different from any laboratory at a German University.

During the following years I became familiar with the laboratory's ways of work and life, and gradually I came to realize how much it all bore the stamp of Sørensen's personality. Not that he intervened directly in the daily work of the assistants and visitors; I was told that he had done so in the past, but that time was over when I came to the laboratory. Now a great part of his day was occupied by work outside the laboratory, by meetings and negotiations. His own scientific work was essentially conducted in collaboration with only one of his assistants, his wife. The rest of us worked rather independently both regarding the choice of problems and the execution of the work, and there were days when we scarcely saw him. But somehow, mysteriously, he always managed to be in the know on his next daily round, and never, never have I seen him with uneasy eyes evading a problem that we presented to him. His ability to penetrate to the core of a matter and his critical faculties were simply fabulous, and never were we permitted to dodge a question that had not been answered to his satisfaction, even if it only concerned some small detail. His uncompromising integrity in such matters was unique.

In this manner he created in his laboratory something which to this day we call »the Sørensen tradition«. We regard this tradition as a precious thing, something which has left its mark upon the work of the Carlsberg Laboratory when that work was at its best, both during the time of Sørensen himself and during the time of his successor, Linderstrøm-Lang. I have been told that during Sørensen's early years his educational methods were more direct, and that he was capable of sternness when confronted with blunders due to stupidity or slovenliness. But by the time I knew him he had become gentle, and our conversations with him were educational simply by the power of his insight and through the unaffected and dignified manner in which the words were uttered.

The Carlsberg Laboratory – a personal view

by *David Glick*

Everyone that I ever heard of who visited the Carlsberg Laboratory as a research worker has been influenced importantly in his professional career by the experience, and had developed personal associations that have survived the erosions of time and distance and have continued to grow deeper and become more enriching. As a holder of a long, maybe the longest, standing record as a foreign visitor, I am honored to write this piece in appreciation of, and with affection for, the Laboratory and the people I have known there.

It is hard to relate my thrill and excitement as a twenty-five-year-old aspiring novice coming to the Laboratory in 1933 to learn about the new development of quantitative histo-cytochemistry that Kaj Linderstrøm-Lang and Heinz Holter had initiated a few years earlier – an elegant technique that permitted biochemical measurements to be made on microsamples of biological material that could be defined histologically or cytologically. I had no inkling on that early October day when I first entered the Laboratory how indelible the mark to be left on my career.

Lang, under whom I had arranged to work, was apparently taken with our adventurous spirit since my wife, Ruth, and I arrived in Copenhagen with only our return-trip tickets and very little money (our annual income was a $1,500 fellowship from the States), and he told us later that Americans travelling abroad were expected to be pretty rich, especially in those times of the great economic depression. When Professor S. P. L. Sørensen and his wife, Margrethe, who also worked in the Laboratory, learned from Lang that we planned to maintain a rather frugal existence in a neighborhood boarding house, they graciously offered us rooms then available in their living wing of the Laboratory, and they would not only accept no rent but not even payment for the cleaning woman. The great man and his wife were also warm human beings.

Sørensen commanded enormous respect from all of us. His prodigious contributions to biochemistry were already well established, and he presented a standard of integrity and dedication that was an impressive example. He was said to have stated that a scientist should obtain at least one reliable new piece of data every working day. I remember his nightly rounds at ten o'clock to check that all was in order before he retired. When I was still working at that hour he always stopped to inquire what the experiment was and to offer a word of encouragement. Sørensen was a model of a man of his time, and a model of a scientist for all time.

Although Lang shared Sørensen's conscientious attention to duty and personal concern for others as well as his scientific excellence, Lang's own talents and bents were many. During my second visit to the Laboratory in 1936–37, he once gave us seminars explaining the relativity theory without the use of higher mathematics. He cared little for the formality so dear to Sørensen and he was unusually unassuming with not much regard for many of the common social trappings. His irreverent remarks about the pompous or uncritically accepted could be quite amusing. He was a fun-loving man with a capacity for boyish glee that, along with an irrepressible joy of living, gave him a charm that was universally engaging. He disliked getting haircuts, and the one tie he wore in the laboratory from Christmas to Christmas usually had the back length dangling much longer than the front. Lang was a »Renaissance Man« whose interests ranged widely and he could do well anything he set his mind or hand to, from the conceptual and the practical laboratory aspects of his science to painting and music.

Like Lang, Holter too had many cultural interests, and he too was entirely unpretentious and shared much of the same outlook on the world as well as in science. Their compatible association was mutually reinforcing and accounted for a great deal of the success of their work together. My professional contacts with Holter were less direct since I worked primarily with Lang, but his help, advice, good company and great humor contributed much to make my visits to the Laboratory more useful and very pleasant. He is the kind of man that wears awfully well, and respect and affection for him continually deepens.

How fortunate I was to be in the Laboratory with these three men – a starting scientist could not have asked for better influences and finer examples of what a scientist and a man should be.

Although I happened to be the only foreign visitor in Sørensen's department at the time of my first trip, when I returned two years later there were a number of us from a variety of countries, including four from the United States. For those of us who were working in histo-cytochemistry, it was a particularly exciting time since it was then, in 1936-37, that Lang invented the additional techniques for micromeasurement of reactions in droplets by dilatometry and in very small volumes by Cartesian diver gasometry.

At that time, too, Lang invented the cryostat for improved microtome-section sampling of fresh-frozen tissue for quantitative histochemical studies. While this was just another instrumental innovation in the development of the technique, it proved to have a much wider application. Now, about forty years later, the cryostat, with a variety of mechanical modifications, has become a common adjunct in perhaps most hospital-pathology and other tissue laboratories all over the world. The commercialization of this instrument, as well as of others originated in the Laboratory, *e.g.*, the constriction pipette, pH-stat, etc., have contributed in their way to the welfare of mankind in the tradition of the »golden words« which are painted on the wall of the staircase leading up from the entrance hall of the Laboratory*. (I have always felt it unfair that the financial gain from inventions from the Laboratory could not return some support to it. Commercialization of the inventions can make their benefits more available to humanity, but this value would not be lessened by suitable arrangements that gave the Laboratory some share in the profit derived.)

A fortunate attitude in the Laboratory was the relative tolerance allowed the researcher to follow new leads in his work, even if the Laboratory were not set up for it and it would require ordering special supplies – within reasonable limits, of course. For instance, when I was measuring the histological distribution of cholinesterase in the stomach, in 1937, I had a particular interest in investigating the substrate specificity as well. The series of choline esters required for the study were not commercially available and had to be synthesized. Since the Laboratory was not really equipped for organic synthesis, Lang let me order the glassware and chemicals needed, and I proceeded with the work. The compounds I had to use have a strong fish smell, and there was a time then when I had the impression that cats were looking at me with more than casual interest as I passed by.

Incidentally, this work led to a midnight commotion in the Laboratory. What happened was, that during the synthesis of one of the choline esters, the reaction did not proceed as expected so I let the mixture stand, hoping it would have reacted by the next morning. As it turned out, the reaction was exothermic and heated up so rapidly once it got started, which was in the middle of the night, that the stopper was blown out of the reaction flask. The next morning I learned that Sørensen received a frantic phone call at midnight from Ø. Winge, the Professor of the Physiological Department who lived in the

* No result of the activities of the institute which is of theoretical or practical importance must be kept secret.

extension of the building nearest my work room, complaining bitterly of the poisonous fish smell coming from the Chemical Department laboratory. Sørensen donned his bathrobe, went down to the laboratory, opened a few strategic windows to dissipate the alarm, and eventually peace, quiet and sleep prevailed once more.

Neither Sørensen nor Winge ever said a word to me about the incident, having passed the matter on through proper channels to the proper subordinate, Lang. However, Lang thought the whole thing was hilarious and both of us had quite a laugh over it.

This happening prompted one of the verses in a poem I composed for a party in New York for Lang in April 1939 on his last visit to the United States before the war:

Glick worked with choline esters pure,
And quite a smell did raise.
His syntheses were met by all
With loads of negative praise.

His favorite dish you'd think was fish,
Or some degenerate stuff,
Or compounds yet unknown to man
Or isomers thereof.

There'll come a day, he'll rue for aye
When on the judgement bench
Saint Peter's gates will firmly close
To keep away the stench.

While all of us in the Laboratory worked hard, there was time for lighter things and a camaraderie among the group that made being there a happy as well as an instructive experience. Lang's caricatures of the laboratory people, sketched on a blackboard, were a delight to all of us, although Sørensen, no

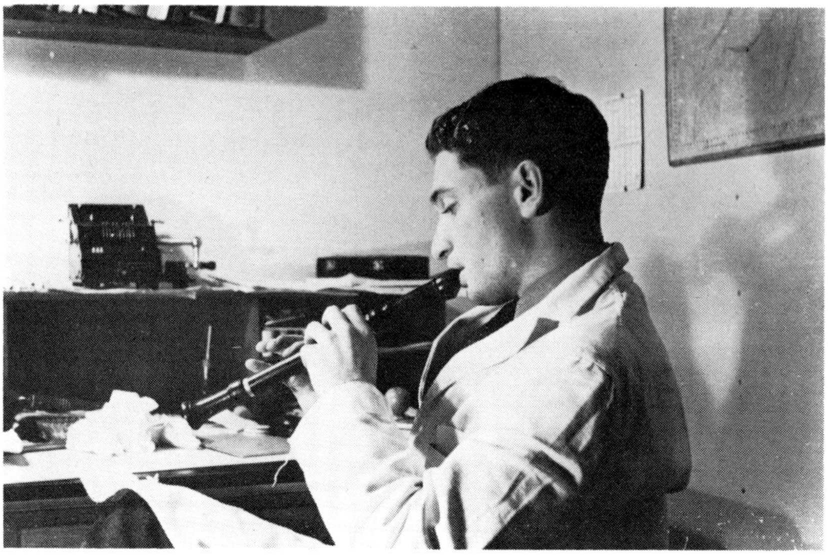

76 Henry Lanz, between experiments.

doubt feeling that this levity was not quite in place in the laboratory, would walk by not letting on that he had seen anything.

Friendly workers from different scientific backgrounds in the same laboratory are bound to teach each other a good deal but, aside from this practical value, the personal relationships developed can be even more important. How to measure the value of friendship – the long winter evenings with Bill Doyle and Henry Lanz when we heavily dented our limited means by indulgence in relatively extravagant Sunday dinners at Frascati's (»The Châteaubriand, please, medium rare, with the Bernaise sauce!«) even though it meant austere living for at least the rest of the week; or the times Lanz played Gershwin on the rented piano in his room and told his tales of undergraduate days at Stanford (I never dreamed that one day I would end up there); or talks with Doyle about Cambridge, where he had spent the previous year, and about books, politics, love, life and all the things »serious« young men discuss; or our happy get-togethers with Lila Miller; or bicycle trips into the country with Rollin Hotchkiss whose dry wit and clear view of the world were an ever-fresh delight; or hikes with Holter guided by his enormously detailed maps of Sealand and with his asides, (»Hm-m-m – that barn should be one the other side of the road – well, you can't have everything!«), and with his great store of Mrs. Pollack and other stories, and the intriguing glimpses into his happier and also more difficult past.

The congenial atmosphere within the Laboratory was maintained in a variety of ways. Lunch-time was livened-up by beer supplied free then, as now, by the Carlsberg brewery, and by contests of skill in tossing beer-bottle caps overhead backward to try to land them on the narrow space of the top shelp on the wall. Then, after some weeks the caps were collected and put into a glass bowl, and a sweepstakes was held to see who could come closest to guessing the number of caps. The winner was given a small prize – all due attention being given to the ceremonious presentation. Another lunch-time feature was shooting »Carlsberg rockets«. The rocket was made by encasing the head of a safety match in metal foil from the inside of a beer bottle cap. It was poised at an angle in the mouth of a bottle and a flame from a match placed under the head set off the rocket in a trail of smoke on an unpredictable trajectory to cries of admiration for the more spectacular flights. Then there was the flow of outrageous observations of Carsten Olsen, (»The opera, Tannhäuser, the best bargain in entertainment – it costs less per hour!«). Another interest at lunch was added by occasional visits of G. Hevesy, whose revolutionary contribution of the isotope tracer technique had just begun then; Lang provided him with some facilities of the Laboratory, he collaborated with Olsen, and Lanz performed some chemical analyses for him.

Of the various parties held, the most elaborate was the annual Christmas party. It was staged in the second-floor meeting room under the portraits of Kjeldahl, Sørensen, E. Chr.

Hansen and other Laboratory notables. The walls were decorated with caricatures of laboratory people that Lang had drawn for the occasion, some of us composed verses to be sung to familiar tunes, and a banquet was served replete with appropriate drinks and wine for each course to the final coffee and cognac or liqueur. Much skåling went on, remembering the penalty – said to be a pair of gloves for any husband who forgot to skål with his wife –, the proper short speeches were made by the chiefs, and there was group singing and general fun. Then the room was cleared for dancing while the older folks repaired elsewhere to play cards and smoke. The laboratory diener, who also served as glassblower and apparatus maker, cleaned up the next day, in part by drinking the remains in all of the bottles of cognac, Cherry Heering, and anything else in sight.

The friendly attachments made between us as visitors and between us and the Laboratory staff, no doubt contributed to the loyalty we developed to the Laboratory itself. After the intervention of the war years, we in the United States learned that the Laboratory had continued to be productive in spite of the trying times, but that certain supplies had run short that were not immediately available, particularly analytical glassware. Through the initiative of Rollin Hotchkiss, a shipment of glassware was sent as explained in his 1945 letter:

> »While Dr. Linderstrøm-Lang was in this country recently, he admitted that his people at the Carlsberg Laboratory were being handicapped in their work because of the unavailability since the war began of transfer pipettes and volumetric flasks to replace normal breakage and to build up their supply. Since these items seemed perhaps more important to him than any others, the fact was mentioned to a number of those over here who had enjoyed the hospitality of the Carlsberg Laboratory in happier years. Contributions began pouring in and we were

77 The Lunchtable, 1951. Names of persons given in chapter 14.

able to purchase the following items: Volumetric bulb pipettes: 12 each – 1, 2, 5, 10, 15, 20, 25 ml. Volumetric flasks with interchangeable ground glass stoppers: 6 each – 10, 25, 50 ml.

These were packed and shipped about Dec. 26, 1945 ... No letter of presentation was sent, but a memorandum of the glassware items and the valuation was mailed, together with a list of the contributors, as follows: George S. Avery, William L. Doyle, David Glick, A. B. Hastings, Heinz Herrmann, Rollin D. Hotchkiss, Hery Lanz, Jr., Milton Levy, Oliver H. Lowry, Gertrude Perlmann, Harold H. Strain.«

In 1949, when my wife, Irena, and I made our first visit to the Laboratory after the war, I remember that we were taken around Copenhagen on the evening of May 5 to see the lighted candles in the windows of the homes commemorating the end of the dark years of the Nazi occupation. To us from the outside, it was a commemoration of the spirit of the Danish people who had remained unbowed and had upheld the principles of decency and of freedom to the admiration and the gratitude of all civilized people.

By this time in 1949, Lang had already withdrawn from the histo-cytochemical activities which were left entirely to Holter's group. The group then included John R. Gregg and Edith J. Krugelis from the United States who, along with Søren Løvtrup, Max Møller, and also Martin Ottesen and others from the Chemistry Department, and Erik Zeuthen from the Biology Institute, added to the list of our continuing Laboratory friendships.

Those of us who came to the Laboratory to learn and apply the histo-cytochemical technique, acted as emissaries to spread its use in our own countries. Through initiation of university courses as well as by the less formal training of fellows and others, and through our own extensions and applications of the technique and the resulting publications, we were responsible in some degree for an international interaction with a phase of the Laboratory activity that was beneficial to all concerned.

Of the histo-cytochemical training and courses offered in the United States that I know about, first there were those of William L. Doyle. He began this instruction in the later 1930s at the Bryn Mawr University Biology Department, and continued it at the University of Chicago Anatomy Department. Subsequently, university instruction was also provided by others, *e.g.*, Oliver H. Lowry at the Washington University Pathology Department. Obviously, department labels have no special significance in this subject fundamental to all biological and medical science.

Extensions of the methodology and technique by previous foreign visitors to the Laboratory took various forms. The ingenious enzymatic cycling technique developed by Lowry, that greatly amplifies the sensitivity of measurement of a wide variety of enzymes and metabolites, was summarized in the book by him and Janet V. Passonneau, »A Flexible System of Enzymatic Analysis,« published in 1972.

My publications on the general technique, including the contributions of all involved, started with, »Methods and Applications of Enzyme Studies in Histological Chemistry by the Linderstrøm-Lang Holter Technic,« published in 1935 in the Journal of Chemical Education, followed four years later in the same journal by, »Recent Advances in Histological Chemistry by the Linderstrøm-Lang Holter Technic and Development of Ultramicro Technics.« Then, after some reviews in annual volumes, the book, »Techniques of Histo- and Cytochemistry,« which was dedicated to Lang and Holter, appeared in 1949. Other relevant writing was the article, »Linderstrøm-Lang, Kaj U., and Holter, Heinz« in the encyclopedia-like book of E. V. Cowdry, »Laboratory Technique in Biology and Medicine,« third edition (1952), with an up-dated version of the same article in the fourth edition (1964) of V.M. Emmel and Cowdry. Finally, »Quantitative Chemical Techniques of Histo- and Cytochemistry«, (Vol. I, 1961; II, 1963) summarized the published material up to that time, and was dedicated to the memory of Lang who had died in 1959. This work carried a Foreword by Holter that gave appropriate tribute to the great abilities and contributions of Lang in this field, contributions that covered only about a fifteen-year period out of his whole career but that have had, and will continue to have, a fundamental influence.

The work of Holter and his Physiological Department established in 1956, following his Cytochemical Division initiated in 1943 within the Chemical Department while Lang was its head, continued the tradition of combining research work with the training of foreign visitors who came steadily from many countries. The laboratory assistance and instruction given them, especially by Karen Holter, and the help and collaboration afforded by Max Møller were greatly appreciated and served to strengthen further the happy bonds generated between the large international community that was formed and the Laboratory. With the retirement of Holter from full-time activity in 1971 and the continuing cytochemical activities by Erik Zeuthen and his group at the Carlsberg Biological Institute, the Lang-Holter phase of the development ended at the Laboratory itself.

While the impact of this quantitative histo-cytochemical development extended internationally, it is neither possible to trace all of the ramifications of its influence throughout the world, nor is it necessary to select either individual places where such impact has been impressed or specific projects that have been evolved. It is enough to say that this influence on the development of all of the life sciences will continue to increase in an ever-widening wave that will reach shores we cannot even surmise, and its significant contribution will continue in the unending quest for means of bettering the human condition through knowledge and its saner uses.

Those of us who have been involved have very special pride for whatever we may have added. Our own careers bear the stamp of, and our lives continue to be enriched by, our experience and our association with the Laboratory and its people, and we shall always have a warm and thankful place in our hearts for both.

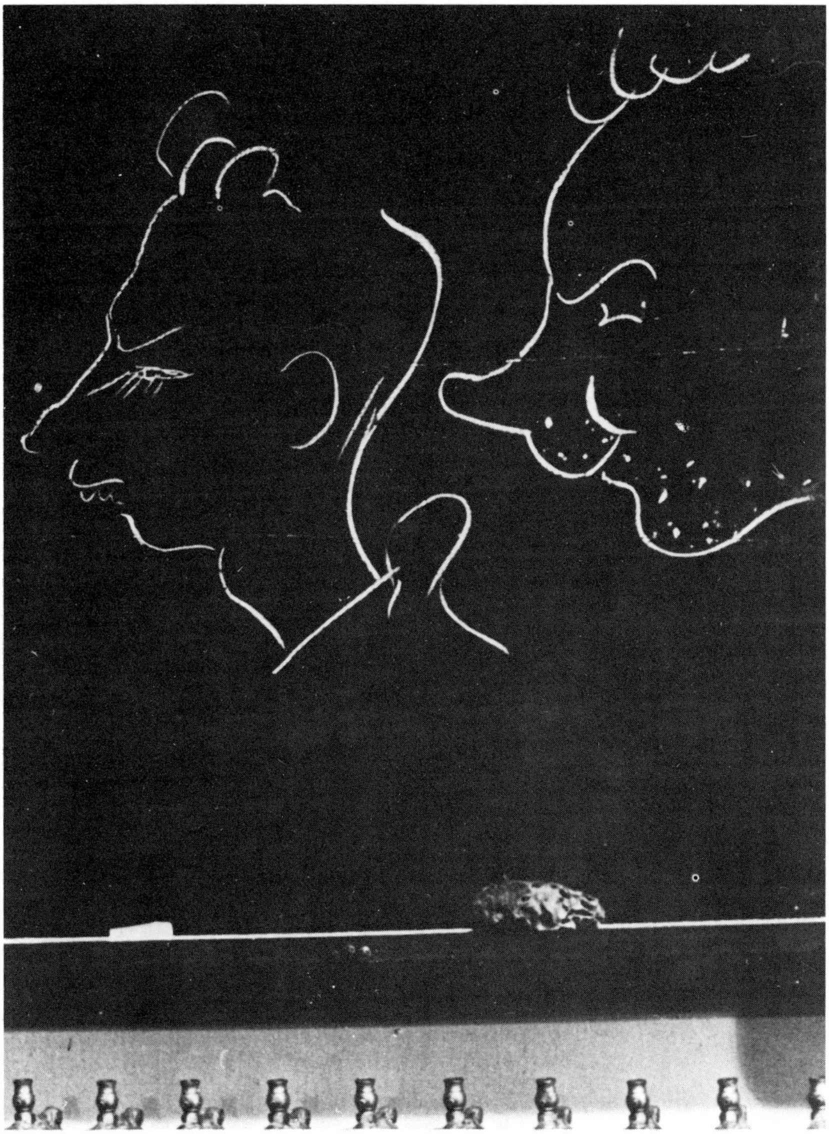

78 What S. P. L. Sørensen politely did not see:
R. D. Hotchkiss and H. Holter, drawn by Linderstrøm-Lang on the blackboard of the laboratory.

Apprenticeship among the stars

by *Rollin D. Hotchkiss*

Those of us who worked at the Carlsberg Laboratory in the 30s and 40s not only acquired precious memories, but were sure to find a more exciting future afterwards. And that future, like the stay at Carlsberg, brought us scientific, social, personal and spiritual values for the whole scientist. Taking the licence of an old-timer – to look back upon a future with hindsight – I would like to paint a few strokes suggesting what it meant to be a Carlsberg alumnus of those stimulating years.

First and foremost there was the invigorating laboratory atmosphere itself. The visiting fellow quickly sensed the implicit ideal of science which prevailed. It was a science of ideas – concepts which help make order in biology and chemistry (»boldness in the face of dogma, humility before fact«). Consequently, what attracted our interest were ideas undergoing refinement or refutation, whether at an early or a late stage – while measurement or empirical observation did not constitute »research problems« (as so often elsewhere) but only respected tools. The interaction of concept and method was so complete that the Laboratory over many years was notable for introducing imaginative methods to improve vastly measurement (pH, solubility of proteins, micromole titration, gasometry, density change and reduced mass) to advance in turn the testing of biochemical ideas.

The young visiting biochemist could not help being moved, to come into the laboratories planned by J. Kjeldahl, who many years before had brought to protein chemistry the dependable and convenient nitrogen analysis bearing his name.

Even more extraordinary was to see in person his successor, S. P. L. Sørensen, the very giant who in the first decade of the century had defined for all time the logarithmic scale which demarcated acidity in units meaningful for enzyme action. How typical of his thoroughness that in one vast fundamental paper he had defined the pH scale, prescribed its precise electrometric measurement, its convenient colorimetric estimation with indicators, and its control with a carefully chosen and defined set of standard buffers!

There was lively and busy coming and going of purposeful workers, but also a basic courtesy toward everyone. Professor Sørensen in his time, passed through the laboratories, morning and afternoon, »making himself available«, greeting everyone but only stopping if the colleague asked for his attention. When he was thus solicited, his eyes would twinkle and he could be courteous and helpful in the extreme: to the point of hunting up special protein preparations, old data, special items of equipment from his own supplies. If his tour through the Laboratory happened to coincide with a rest moment when a couple of workers were practicing upon the recorder or laughing over anecdotes, the Professor would politely look in the other direction, and his pointed beard would look very elegant and trim indeed.

Kaj Linderstrøm-Lang, as Sørensen's successor (installed officially during 1937–38, the year I was there), achieved the same availability in his informal way – in an unbuttoned grey suit hanging over his frame, with a nearly consumed cigar, and a ready smile which would become boyish and suffused with a bright-pink blush, under the embarrassment of once more being asked to behave like an authority. Because of his brilliance and kindness he was asked many questions, and stopped often and long, so his rounds were inspirational rather than regular. Often he would volunteer interesting news and information to us. Once I asked him if he thought the iodoacetate I was using changed in pK when it reacted with protein sulfhydryl groups, buffering some of the protons (HI) it released. He thought so, and discussed it; two days later he produced several pages of carefully argued text and calculations, in English, showing how to estimate the error and moreover, the instantaneous change in carbon dioxide tension as I mixed (from a Warburg vessel sidearm) a bicarbonate solution not containing protein with one which did. This led me to switch to iodoacetamide and I went on to develop a gasometric method for determining sulfhydryl compounds, including proteins.

All such magnificent help was offered modestly; in fact none of the Carlsberg people would make an inquirer feel ignorant or put down – and they certainly helped cure their visitors of any such academic practice. Indeed, a reckless or foolish question was so generously enjoyed that we all, from the professor down, were likely to tell and retell our »boners« for the good fun they created.

Heinz Holter had most of his days planned with time-controlled experiments. Yet he always could stop for a question, and if he deferred answer until later, he was faithful and reliable in returning to deal with it. His patience and courtesy toward the logistic and psychologistic problems of any earnest visitor was so outstanding that it was difficult to realize that for years his own status in the official scheme was relatively insecure. He was a storehouse of knowledge of cellular enzyme biochemistry and of how to do or obtain things for investigating them.

In this environment one expanded to near his capacity and began to grow. Since the environment made a vivid impression, it was remembered and influential throughout one's life.

The large, high-ceilinged laboratory rooms were somehow spacious and cheerful even though the dark cabinets and tables seemed to be filled with the residuals of earlier projects: titration devices, special balances, various baths and heaters adapted uniquely to special ends. The faithful glassware washer returning washed pipettes placed the special sizes (such as

3 and 4 ml sizes, the 0.5 and 7.5 ml ones) for years in drawers near where their most constant users had once worked. Thus, one sensed the history and in some degree retraced about the laboratory the steps of long departed predecessors.

The visitors beginning to come in larger numbers around the early 30s might arrive knowing very little Danish, but they found more than half of the people at the Laboratory able to speak English, German or French, and many of them eager to become more practiced at it. Thus, if the visitor were not boldly insistent, a Danish which might suffice for shopping or bargaining in the street or making reservations by telephone or at the railroad, would hardly be developed further in the Laboratory. But Danish customs were there to be embraced freely, or ignored according to taste; all could be forgiven in the foreigners. One quickly enjoyed the community of the common lunch table with its discussion and fun; the wit and laughter as likely to come from one end of the table as the other, and to arise out of any topic within the range of experience or fantasy represented by the group. In reasonable weather this event was likely to be terminated by the ringing call, »Frokost-tur!« by Holter – announcing an after-lunch walk along Valby streets, during which one-to-one conversations and friendships often arose. One was delighted to find that on an anniversary, the celebrant himself proclaimed his joy by providing for the staff the delicious Danish specialty – called »Viennese bread« in Denmark where they are at their best, and »Danish pastry« in other places, where they suffer greatly in the translation. The lucky donor's reward was the warmly repeated »til lykke!« wishes from his colleagues.

When I arrived in July 1937, M. Levy, L. Weil and F. Duspiva of the new wave of cytochemically-directed visitors had already come and gone. Present at the time were D. Glick, H. Lanz, W. Doyle, J. Lehmann, K. Mogensen, and soon to come were H. H. Strain, G. Avery and M. Sreenivasaya (who later imported his own assistant to prepare meatless meals and tireless experiments). Representing the protein chemistry interest of the Laboratory there had recently been A. H. Palmer and A. H. Johnson, and there now were (besides G. Haugaard and C. Olsen of the staff) A. Grönwall and G. Johansen. The Physiological Department headede by Ø. Winge operated in a rather similar way, but exchanges with the outside were less active, and between the departments were mostly limited to infrequent cooperative work and the seminars and special occasions.

The seminars attracted a good many scientists from other laboratories in Copenhagen. Often seen were Gertrude Perlmann, Heinz Herrmann and Fritz Lipmann from Carlsbergfondets Biologiske Institut, and Herman Kalckar from the University, who became friends of many of us then and later when they came to the United States. I have been assured that I contributed atmosphere if nothing else when »talking« with the last two mentioned – as we mused together, Lipmann peering smilingly toward the heavens, Kalckar darting piercing looks this way and that but with eyes almost closed, and I gazing down at my feet; language was no problem, because the words fell almost unheard at the rate of five or ten per minute.

It would take too long to tell how individuals within and without the Laboratory so readily became friends and enriched each other. Yet it is a pity that the scholarly science historians making their sober records are likely to overlook the tremendous focal, stimulating and influential effect of a particular laboratory environment upon young workers, or the effect of these workers upon each other. The tangible joint publications which of course are faithfully noted, reflect so little of a world in which the vital connections are often intangible! (Just as these historians usually assume that the earlier publications must have influenced those who published later, although a vital flow of stimulus or information had often gone earlier in the opposite direction.)

The earnest discussion of science, or equally, the pleasure in each other's company, could spread from the Laboratory to other places and times. The after-lunch walks led to dinners, or the weekends brought bicycle trips (David Glick and I covered a considerable area of Denmark together during the three months we overlapped in 1937) or walking trips with Heinz Holter among the forests, dells and marshes of Sjelland, like the other migratory birds. (Heinz once looked up from his detailed map to point out a mockingbird in the tree above; we immediately crowded round to see what kind of symbol was used in the map for mockingbirds.) The anniversaries or holiday parties brought out the fun of community talk or singing, often enriched with violin-piano numbers or Swedish songs by Lang and his friends.

The Carlsberg fellowship indeed extended round the world and into later years. The mere mention of a stay at the Laboratory would bring strangers together without regard to their specific academic interests, but they were often not really strangers for the »family news« often traveled ahead of the person. When Lang visited New York shortly before World War II, a score of former alumni got together, some for the first time, to greet him with our own songs and festivity. Some exaggeration may have occurred, especially with respect to a quiet member like Heinz Herrmann, for whom we sang:

'Twas a cinch for young Herrmann the Tall
Analyzing a sample quite small
He did earnestly seek
To improve the technique
Not to need any sample at all.

Likewise, A. H. Palmer, whose beta-lactoglobulin preparation by dialysis was used much in the Laboratory, was imagined as creator of a stunning experiment:

When Palmer, our great crystallizer,
Put a cow in a big dialyzer
Her milk came in fistfuls
Of big shining crystals
You can well see how that would surprise her!

A little later, Paul Zamecnik and Chris Anfinsen and their wives, whose stays at Carlsberg were abruptly terminated by the German invasion of Denmark in 1940, suddenly turned up in New York, bringing home to us first-hand views and evidences of the overgrown ambition of the Nazis, which was to hang darkly over our friends and ruin so many lives before it could be stemmed. But at that first meeting with new friendships forming, the preposterous did not yet seem true, and we were happy with mutual pledges and warm assurances and recommendations from the old friends back at Carlsberg. When, some thirty years later, we congratulated Chris Anfinsen about his »message from Stockholm« (the Nobel Prize), he referred to Carlsberg and Lang: »thank you for the warm note ... Kaj's old boys have a way of shoring one another up and I think he will be walking down the aisle with me, stepping sharply on my toes! Love to you both...«

How did so much that was good come to a peak at the Carlsberg Laboratory? How can this good be preserved, and transplanted and propagated? I can offer only my impressions, based partly upon my observation of all of the research institutes in which I have been fortunate enough to work. Important, I think, was the overriding sense that science and clear thinking are not only important but enjoyable. At the Rockefeller Institute where I have spent so many years, there was ample appreciation that science is more important than the individual, but only at choice times and places within it was the pleasure of doing science freely admitted (at its best, the laboratory of O. T. Avery could supply this, so did that of K. R. Porter, and there were others). At the Cold Spring Harbor

Laboratory under M. Demerec, the sense of enjoyment was often manifested, and with some exceptions, the individual did not seem to overshadow the science. In such places, great steps could often be made.

Also important is an atmosphere in which rank and individual prestige are not very important. An »open« situation with visitors coming and going at intervals allows contribution by all without too much concern for their relative competitive positions. This makes a good »growth situation«. The Carlsberg Laboratory and Cold Spring Harbor could provide this at their best periods, partly because of their nature, organization and direction. It did not depend primarily on the magnitude of their financial support, which was not very great, particularly at Cold Spring Harbor. (After a productive organization is created, giving more support without new conditions can widen its beneficial influence, but at times also stimulate intramural problems and downfall.) It depends more upon the liberality and quality of principles upon which the support is used. Great places like the Rockefeller or Pasteur Institutes seldom, in my opinion, could achieve this optimal openness as a whole, but sometimes they did in spots, for a time only.

The whole success, of course, depends greatly upon the generosity and flexibility of mind of able directors. The obligations imposed on science administrators, foundation heads, and their assistants, tend to make them crave and seek *predictability*. They may think sometimes in terms of »this many units of developmental biology, that many units of biochemistry« and so on – the units being variously dollars, assigned space, or predictable scholars. It seems almost to be required of administrators and of the programs which they are mandated to further. I think the Carlsberg Laboratory benefited during three decades from an enlightened Carlsberg Foundation, some of that time from an imaginative Rockefeller Foundation (Natural Sciences Division) and from a generally intellectual and liberal Danish cultural environment. But, best of all, it was guided by scientists of the caliber of Sørensen, Linderstrøm-Lang and Holter, intelligent and vigorous themselves, and possessors of strong viewpoints, but modest enough to give help to, and place trust in, a host of younger people who crowded around to follow science under their influence. My best wish and hope for the Carlsberg Laboratory on this anniversary is that its sponsors have the will and insight to strive ever to maintain these conditions favorable to science and human development.

Carlsberg Laboratory, Chemical Department – late forties

by *L. Vandendriessche*

We arrived – my wife and I – at the Carlsberg Laboratory on a Saturday morning in October 1946. The Danish government had at that time provided grants for young scientists in the making. I obtained one, and we intended to stay for three months in this unknown laboratory and this unknown country, driven more by an urge to go to a foreign place after all those long war years of seclusion than by the desire to work and study in a definite institution. Another country with grants and another laboratory would have been accepted with equal enthusiasm.

The reception was unexpected and overwhelming. While we were waiting in the big entrance hall, Linderstrøm-Lang arrived, then nearly fifty. I need not describe him, in his white lab coat, with his blonde hair and very young boyish face, which from the first moment established a reassuring contact that inexplicably commanded admiration and respect. We came from a country where we still had ration cards – we were invited to lunch. Nothing seemed scarce here; there was milk, butter, cakes and the Carlsberg beer, which we met for the first time. But above all there were those present: besides Lang, Heinz and Karen Holter, C. F. Jacobsen, Carsten Olsen, Gordon Johansen, amusing, teasing, but friendly, attacking us with all kinds of questions about our plans, our country, and about the war.

Finding a place to live was a difficult problem at that time.

The laboratory had found us a room in a pension mostly inhabited by old people. It was situated on the other side of the town, a long streetcar-journey away from the laboratory. It was not terribly clean (the menu of the week could be read on the apron of the landlady, as Lang remarked) and the food was very different from what we were accustomed to and not too good. But it had one big advantage: it was cheap.

And so a new life started for us, for me at the Carlsberg Laboratory, for my wife in Rigshospitalet (the University Hospital) and the Finsen Institute, and it has determined the pattern of the rest of our lives.

Denmark and the Carlsberg Laboratory had a great attraction for Belgians. It was not long before three other Belgians had arrived (H. Chantrenne, E.P.O. Fredericq and J. Leonis) all unknown to us. Later a Norwegian (N. Eeg-Larsen) came, followed by some British and the American invasion.

The atmosphere of the laboratory had the same effect on all of them as on us. After some weeks we started to talk about how nice it would be to stay longer than intended, and this talk was followed by heroic endeavours to obtain permission from home to stay away longer, and to find the necessary money. Most of us succeeded in making the three months last six, twelve, and (in our case) even thirty. And when finally it had to end, there came a longing to return. The Carlsberg was the place of which we dreamed and talked. At congresses, Carlsberg alumni evenings were organized, good friends were persuaded to go to that wonderful place, and later, when we had our own laboratories, we sent our pupils to the Carlsberg Laboratory to get the same training and experience.

What was so special at this period about this place, and what was the reason for its lasting attraction to all who had landed there?

Circumstances were of course very special. We were all very young. We all ended up by living at that same pension. We all had to share rooms, have meals together, and ride together in

79 Carlsberg reunion in Brussels, 1955 (Names, see chapter 14).

the icy cold streetcars in the winter of 1946–47. We spent our days discussing, mostly politics in those post-war years, but also our work and science in general, and we learned much from each other since we had been trained in a number of different ways. We were lucky enough to have no serious incompatibilities, and there was no reason for competition, so we all became, and still are, good friends. Then there was the laboratory, with its very special atmosphere of the post-war years. Everybody seemed on the move. All Europeans wanted to go abroad, and for many of them the goal was America. Then the Americans came over to visit European laboratories. Many of them went by sea to Sweden, and this often meant passing through Copenhagen and, for a biochemist, visiting the Carlsberg Laboratory. In certain periods we had one such visitor nearly every day. They were invited to lunch and most of them gave a seminar about their work. But this was not the most important aspect. The discussions about everything, politics, music, life during the war, a newly published book, the atom bomb, always with fresh people, each with his own interests, were for many of us something completely new. After all, we had spent years in small unimportant laboratories, secluded during all those tragic years, surrounded by material problems of food and security. And now all this seemed over and a new unknown world opened up.

But above all, there was the work in the laboratory. Equipment was not so exceptional according to modern standards, but this was more than made up for by the inventiveness and the formidable fundamental knowledge of Linderstrøm-Lang. You could always go and see him, or see Holter or Jacobsen. You could have discussions with them – not necessarily very serious or long – and learn from them how to think about a problem, how to distinguish between fundamentals and details, how to avoid useless and futile experiments. And there was the assurance that, when you ran into real trouble with your work, someone would help you to find a way out and an explanation.

Then there were the evenings. We had moved from our distant pension and had come to live near the laboratory, again many together in the same house or close to each other, but now in places where there was no dinner. So we got permission to cook our own meals in the laboratory, which was very cheap and very time-saving. Work continued afterwards, in the quiet relaxed evening atmosphere, with Lang or some-one else coming over for a chat, or with music between experiments. Here again the most rewarding thing was that we were all together during meals, endlessly discussing science and politics, our Norwegian telling about his life in the concentration camps, our American explaining to us, according to »Time Magazine« (which arrived on Friday) the dangers of Russian communism, and we, leftists as most of us were, subverting him, until the next »Time« arrived and shored him up again. There were the invitations to Lang's home, and the musical evenings, where some of us had formed a small chamber orchestra. But there were also the dinners in honour of some distinguished guest to which we were sometimes invited. Or just evenings with him, or with the Holters, in houses full of books and music, together with people from other laboratories, such as Niels Bohr's or August Krogh's institutes. On Sundays we used to walk with the Holters in the Danish countryside, or sail out to sea with them.

Then there was the highlight of the year: the laboratory Christmas party, preceded by feverish days. Every newcomer had to go to Lang's office to pose for his portrait, and Lang himself was in a nervous and excited mood, doing his drawings and composing his Christmas song about life in the laboratory. Then came the evening, all together in the big room upstairs, which at that time normally was used only for meetings and lectures, but that evening was full of an undescribable atmosphere that one must have experienced to know what it can mean. I remember that my third Christmas party was the definite end of my much prolonged stay. I remember saying to Lang, and repeating: I will come back, I will come back – and so I left. I did come back, as many of us did. I came back many times, meeting new people, seeing the laboratory change, seeing people change, seeing people go away to other places, and seeing a page of Carlsberg's history end with Lang's death.

It was by sheer luck that we came to the Carlsberg, and luck it turned out to be. »The best years of our life« – the best and the most rewarding. – Thank you, Gamle Carlsbergvej 10!

80 L. Vandendriessche, drawn by Linderstrøm-Lang for the Christmas party 1947.

Recollections of Carlsberg Colloquia 1934–38

by *H. M. Kalckar*

As far as I can recall, the Carlsberg Laboratory Monday afternoon weekly colloquia were the only regular weekly workshops in biochemistry in Copenhagen at that time. We all cycled in rain, drizzle or sunshine up the hill (Valby Bakke) to the »Parnassus«, the Carlsberg Laboratory.

Old S. P. L. Sørensen was the gracious chairman of the colloquia. At that time many of the workshop talks were somewhat outside his scope. Nevertheless, he had a happy way of getting the discussion started with simple, perhaps naïve questions which paved the way for making the audience feel a little more at ease. In later years, he used just to thank the speaker, adding – »these matters are indeed interesting but also very complicated«, which was perfectly correct, and once more was an invitation to the audience to feel free to discuss.

Sørensen and Linderstrøm-Lang created a free and easy atmosphere at these colloquia, quite different from most university lectures or the weekly lectures which I later encountered at other successful institutes.

These were the years when Linderstrøm-Lang conceived some of his new ideas about denaturation of proteins and the initiation of proteolytic activity.

I remember a lecture in 1938 by Lang on some of these concepts, (formulated with Rollin Hotchkiss and Gordon Johansen) in which in an almost prophetic way he warned against too facile extrapolations from data on proteolytic splitting of synthetic peptides to the structure of native proteins. He

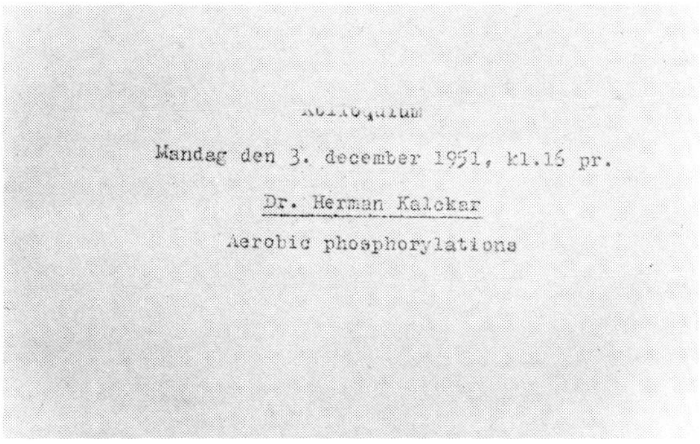

Mandag den 3. december 1951, kl.16 pr.

Dr. Herman Kalckar

Aerobic phosphorylations

81 How the colloquia were advertised. The Carlsberg Laboratory is not mentioned; everybody knew the small yellow sheets of foolscap, and where the colloquia were held.

was influenced here by the views of M. L. Anson and A. Mirsky (two frequent guests of the Laboratory) on the reversibility of the denaturation of proteins. Assuming an equilibrium between genuine (native) and denatured protein, Lang and Hotchkiss suggested that only the denatured form contains peptide bonds susceptible to fission by trypsin. Removal of the denatured form by hydrolysis forces the process from the native to the denaturated form strongly in the direction of the latter. This cautious formulation was expressed by Lang as a »warning against the conclusions that genuine proteins contain (hydrolysable) peptide bonds because they are split by proteinases like trypsin. They give an indication that peptide bonds are formed or »appear« (like SH groups) upon denaturation, but they are not conclusive enough to decide whether or not some hydrolysable peptide bonds are preformed in the molecules of the genuine globular proteins.« This was a close forerunner to Lang's later concepts about secondary and tertiary structures and also to L. Pauling's concept of the hydrogen bond, later L. Pauling & R. B. Corey's α-helix structure, and Lang's studies on the fast and slow exchange of hydrogen ions in native proteins.

Around 1936–37, F. Lipmann had further developed his unconventional and exciting ideas about bioenergetics and the role of phosphate in oxidative decarboxylations, as studied by him in Lactobacillus Delbrückii.

The problem of protein synthesis and bioenergetics interested Lang very much. At that time H. Borsook and G. Keighley had been discussing turnover rates of tissue proteins and gross metabolic kinetics and had made themselves spokesmen for ideas which became forerunners of R. Schoenheimer's and H. H. Ussing & G. Hevesy's isotope experiments on the metabolic flux of tissue constituents. We worried about the »expenses« involved in such a continuous synthesis, breakdown and resynthesis of peptide linkages. Lang advanced a theory about »amino aldehydes« as precursors of peptide links, much influenced of course by Lipmann's ideas about energy conservation, forerunners of his concept of acylphosphate, developed by him and by Otto Warburg and his group in early 1939.

Oxidative phosphorylation was briefly presented by H. M. Kalckar in connection with active transport systems.

Lang shared with Lipmann a special admiration for Einar Lundsgaard's new formulation of bioenergetics of muscle contraction and the simplification of the formulation of the Pasteur effect. I recall Lundsgaard's participation in some of these colloquia, especially when Lipmann spoke about sulfhydryl groups and the Pasteur effect.

At these meetings it was also quite natural to speak about yeast biochemistry and physiology, since yeast metabolism had revolutionized muscle biochemistry in the twenties and thirties. I recall a lecture by John Runnström about coupling of oxidation-reduction and phosphorylation, Holger Jørgensen and Jørgen Lehman on oxidation-reduction potentials, and Detlev Müller on the fumarate system in bacteria and fungi. Henrik Dam spoke about factors in blood coagulation, a forerunner to his great work on vitamin K. I do not recall Winge lecturing about his revolutionary work on mating characters in yeast. I may have left for the USA at that time. However, in C. van Niel's second microbiology course at Pacific Grove, July-August 1939 (in which I participated) he told us about Winge's observations.

The Carlsberg Laboratory – A challenge to the M. D.

by *Jørgen Kieler.*

The centenary of the Carlsberg Foundation will give previous visitors to the Carlsberg Laboratory an opportunity to reconsider what they owe this remarkable center of research and what it means to them to-day.

The long list of visitors is dominated by highly qualified research workers devoted to basic science, but also the servants of Asklepios were frequently attracted to this outstanding research institute, which in many ways differed fundamentally from the hospital or university environment to which they were accustomed. It is of my medical colleagues that I am mainly thinking in recounting my own experiences as an example.

Why did they come?

The scientific achievements of the staff and the high reputation of its leaders have undoubtedly played a decisive role in all cases, and in answer to my question many of my colleagues would probably also refer to their interest in a specific scientific problem, suited to a doctoral thesis. However, I am afraid that I have to admit that it was primarily a mixture of curiosity and challenge which brought me into the inspiring sphere and influence of Heinz Holter and Linderstrøm-Lang.

In order to explain this I shall briefly summarize a few personal facts. The second world war left me with much too much experience in the physical and mental sequelae of semistarvation. It was soon clarified that protein deficiency was the most important single factor in the development of hunger cachexia in European concentration camps. This led me to a study of the influence of amino-acids on mitosis and cell growth in vitro, a field of research which had already been opened up by Albert Fischer at the Biological Institute of the Carlsberg Foundation, and to which W. Earle and H. Eagle and their collaborators later made such important contributions.

It was Heinz Holter who was appointed by the university of Århus to make a critical evaluation of my modest contribution, which also was my doctoral thesis. Nobody will be surprised to hear that Heinz gave me and the audience a brilliant lecture in cell physiology, which left me with a feeling not only of admiration and inspiration, but also of curiosity and challenge.

This was my first meeting with the contrast between basic and applied science, and even if I did not fully appreciate the issue I was facing, I felt its importance so strongly that a sabbatical year at the Cytochemical Division of the Chemical Department at Carlsberg became a condition for accepting the position as division head at the Fibiger Laboratory which I had been offered.

I was facing a choice between a clinical career and basic science. It was difficult for me to make up my mind, but I hoped to find the answer at the Carlsberg Laboratory. I did. This is why I came, and this is perhaps the most important thing I owe this laboratory, where I spent one of my happiest years.

It was a strange world I entered in March 1955 when I appeared at my new place of work. I started in the main building and looked respectfully at the portraits of the giants, admired the impressive staircase, and made a solemn promise to myself always to remember and respect the moral codex written on the wall.

Then I found my way to the cytochemical laboratory in the old villa at Gl. Carlsbergvej Nr. 8. It was less impressive. I did not know whether I was expected at 8 or 9 o'clock, but I did not want to be late. I did not know the customs of the place, but I was convinced that I would find all virtues at the Carlsberg Laboratory, including early-rising and precision. I was soon to learn that I was not expected at all.

It was 8 a.m. and I had the whole laboratory to myself. A quick inspection of the rooms told me that fame does not necessarily require luxury. A notice on the wall made a special impression on me. It stated that scientists were entitled to 4 weeks of vacation, technicians to 3 and laboratory stewards to 2. Apparently it was not only the house which was old.

In spite of a vacation period limited to 2 weeks, Mr. Christensen was the first to appear, but before 10 o'clock the laboratory was full of so many busy people that I really could not see how they would squeeze me in. Surprisingly enough it did not cause any alarm when I told them that I was going to work there. Heinz arrived half an hour later, very apologetic, and assuring me that this was unusually late. Although I was a little worried about the lack of space, I felt at the same time relieved. Even if they were admirable, these Carlsberg people, they did not possess all the virtues. That gave some hope for me.

Heinz took me under his wing. He refused to acknowledge any space problems. »We have plenty of room, but we need a little reorganization«, he said. His wife took care of the reorganization, and in half an hour I was well installed at a corner of a desk and with ¼ m^2 of working space.

What followed was a life of which not only I, but numerous colleagues and friends, have the happiest memories. Karen Holter, Birte and Max Møller, and »little« Karen (the secretary) soon made me feel at home, and together with many visitors from abroad they turned daily life into an inspiring, gay, and creative existence under Heinz's kind supervision. The close contact with the Chemical Department gave me the added opportunity of enjoying Linderstrøm-Lang's numerous talents, cheerfulness and eventually also his friendship.

I soon found myself engaged in a study of the influence of pCO_2 and pO_2 on cellular respiration. Karen was the patient teacher who taught me the Cartesian diver technique, and how to make micropipettes and constriction pipettes. Mr. Christensen built the water thermostat, and technical improvements were continuously discussed with Heinz and Max. During these activities I got acquainted with the fun of experimental work in a way which previous tissue culture studies had never revealed. In addition the most expert advice was continuously offered by Heinz and others. Everybody showed an interest in the work carried out by others in the laboratory. Nobody remained isolated.

They were busy days, but not too busy for fun. The lunch hours at the main laboratory became the high-water mark of the day, which even the most modest occasion could turn into gay celebrations, during which Linderstrøm-Lang displayed his unusual social talents.

But the Carlsberg way of life also gave visitors ample opportunity to establish more serious and personal contacts. Here the picnics of the Cytochemical Division will be remembered by many as one of the best Carlsberg traditions. Whenever I drive south through the woods around Gisselfeld and Bregentved I think of Heinz, who during a picnic suddenly stopped me and said: »Look, this is the prettiest sight in Danish nature«. He was pointing to a grass snake hurrying into a pond which it crossed at an amazing speed. The enthusiastic expression in his eyes revealed his delight in nature and natural beauty.

On such occasions the conversation often became more personal. I remember that I asked him whether he did not miss his Austrian mountains, and I believe that I understood him very well when he answered: »Yes I did, until I got a boat and started to sail«.

Similarly, whenever I get in the neighbourhood of Jægerspris I cannot help thinking of Karen, who on our mushroom picnics was the specialist to whom we would turn to hear whether our harvest was edible. Karen looked at the mushrooms,

82 Mushroom hunting in Jægerspris forest. (Names, see chapter 14).

and then she took one and chewed it. After some thoughtful chewing she would pass her verdict: »Very good; somewhat bitter, but not poisonous; slightly poisonous; or lethal.« Then she would smile and look very happy, while she continued to chew her lethal mushroom.

Yes, these were happy days. A life among people for whom natural beauty, music, literature and art were necessities, and for whom the exploration of nature was both a hobby and an important contribution to civilization. In this environment science became an essential part of human culture, and nobody asked about its usefulness. Politics were not much discussed, and the anti-science movement had not yet started.

My visit to the Carlsberg Laboratory helped me to make my choice between a clinical career and basic science. The laboratory and not the clinic became my future. However, this decision was still not easy, and it left some unsolved problems.

Against a background of experience gained in concentration camps, and later in hospital wards filled with patients in the late phases of cancer, I could not help feeling that I was joining a privileged group, to whom the austerity of life had not yet been revealed. Was I retiring to a retreat, or was it possible to accept the challenge from basic science and applied science at the same time?

It was at such moments of doubt that I discussed with Heinz the problems which the scientist has to face sooner or later when he reaches the stage which Heinz described as the scientific climacterium. It was also at such moments that Linderstrøm-Lang revealed to me that he would gladly have accepted some teaching obligations. However, it was during my long conversations with Max that I reached the conclusion that basic and applied science are not necessarily in contrast to one another, and that the gap which sometimes appears between them can be bridged.

This brings me to my last question: What does the Carlsberg Laboratory mean to me to-day?

In trying to answer this question I would like to give prominence to the continuous inspiration and encouragement which I derive from my recollections of the time I spent in the cytochemical laboratory–inspiration and encouragement which is needed, when the social relevance of science is discussed, when financial priorities are to be decided, and when quackery is supported by some superstitious politicians and an uneducated press.

May the Carlsberg Laboratory continue to serve these functions.

IX

The Carlsberg Laboratory and the Exploration of the Sea

by *E. Steemann Nielsen*

The Carlsberg Laboratory was for more than a quarter of a century one of the leading oceanographic centers of the world. It started in a modest way around the turn of the century, when Martin Knudsen, the Danish physicist and hydrographer, allied himself with S. P. L. Sørensen, whose talents for exact quantitative analysis were used in establishing standard methods for the salt analysis of sea-water, later on supplemented by measurements of hydrogen ion concentration in sea-water samples, carried out by Sv. Palitzsch (1911). A mixture of pure salt solutions, called »Standard sea-water« is the fruit of this collaboration, and it is available to this day.

During the same time another Danish scientist, the zoologist C. G. Johs. Petersen, played a great role in the development of Danish oceanography. He was the originator of the quantitative bottom sampling technique and his insistence on regarding marine biology as a science succeptible to quantitative treatment has left its mark in the work of his pupils, one of which was Johs. Schmidt. Both Martin Knudsen and C. G. J. Petersen were very active promoters of the International Council for the Exploration of the Sea which was established in 1902 and which to this day has its headquarters in Copenhagen.

The salt analyses of sea-water were taken up a decade later, when R. Koefoed, a former assistant of J. Kjeldahl, in 1911–12 participated in a German expedition to the Dead Sea and brought home with him samples of Dead Sea water. These samples were analysed by G. Haugaard in the Chemical Department of the Carlsberg Laboratory and the results were published in 1926.

But the real engagement of the Carlsberg Laboratory in oceanographic research began in 1910, when Johannes Schmidt was appointed head of the Physiological Department and ended in 1933 due to his early death. The Carlsberg Foundation did not stop its support of oceanographic activities at this time; an additional 40 years was spent working up the immense scientific material collected all over the world as a result of Schmidt's activities.

Johannes Schmidt was only 33 years old when he was appointed director of the Physiological Department in 1910 but he was by no means an unknown scientist. He started as a botanist, had published a series of botanical papers, e.g. about the flora of Thailand. However, the direction of his career changed in 1901, when he was appointed as assistant to the prominent fishery biologist, C. G. Johannes Petersen, the director of the Danish Biological Station. During cruises with Petersen in 1901 and 1902, he learned not only the essentials of fishery investigations, but also the more fundamental approach to this subject based on an initial investigation of hydrography.

Two years later, Schmidt was appointed as leader of the Danish Fishery Investigations around Iceland and the Faroe Islands. After the establishment of the International Council for the Exploration of the Sea, the Danish Fishery authorities had bought a trawler »Thor«. Every summer for several years, Schmidt cruised around Iceland and the Faroe Islands investigating the spawning of the economically most important fish species, including the fate of their fry. This was pioneer work of great importance for the fishery industry.

83 Johs. Schmidt meets in 1929 on the island of Koh Chang (Thailand) his old collaborator Nai Sug, who 30 years earlier on Schmidt's very first expedition had assisted him in his studies of the Thai flora.

During the second cruise, in 1904, a single larva of the freshwater eel was caught on the »Thor«. This was the first eel larva, called Leptocephalus brevirostris, which had ever been caught in the Atlantic. Some years before, eel larvae had been found in the Mediterranean. This first Atlantic eel larva started the long search for the breeding place of the eel. At this time, no one could have imagined how long and difficult the search would be.

In the next years, Schmidt fished many eel larvae in the Atlantic waters along the west coast of Europe but all the larvae were in a late stage, just prior to their transformation into elvers. Since Leptocephalus brevirostris had first been found in the Mediterranean, Schmidt decided that the breeding place might be there. He realized that only by finding smaller and smaller larvae in their earliest stages would he be able to trace the spawning ground.

In the winter of 1908–09, he went off with the »Thor« to the Mediterranean. The ship was available then as no fishery investigations were carried out in the north Atlantic during this time of year. His primary support for this voyage came from the Carlsberg Foundation. Although he had assumed that winter was the breeding season of the eel, no small larvae were found.

During the summer of 1910, Schmidt again went to the Mediterranean with the »Thor«. It was in this year that he was appointed head of the Physiological Department of the Carlsberg Laboratory. The expedition of 1910 covered the entire Mediterranean. It could now be stated that the frequency of eel larvae decreased from west to east whereas the size increased. This showed that the eel larvae originated in the Atlantic.

Despite its failure to find the breeding place of the eel, the Mediterranean expedition was by no means unsuccessful; just the opposite. Schmidt never expended all his effort on a single, narrow goal. Besides carrying out a thorough investigation of Mediterranean hydrography, an immense number of biological specimens were collected, principally of plankton and fish larvae. The two Danish expeditions had increased the general knowledge of both the physical oceanography and pelagic biology of the Mediterranean to a high degree. The Carlsberg Foundation financed the publication of 11 volumes containing nearly 3000 quarto pages.

Despite his involvement in many other activities at the Carlsberg Laboratory, Schmidt was not diverted from his oceanographic work and in particular, the search for the breeding place of the European eel was by no means given up. He continued at this time as a member of the International Council for the Exploration of the Sea and as leader of the Danish investigations in the northeast Atlantic. Working facilities for his assistants were provided in the Carlsberg laboratory. On the other hand, the next steps in his investigation of the breeding place of the eel were not financed by the Carlsberg Foundation. With the aid of several captains of merchant ships, a Danish man-of-war, and a small schooner, »Margrethe«, hired by the Danish government, large parts of the north Atlantic were investigated. The materials collected during this wide search showed that the breeding place of the European eel was in the western part of the Atlantic, specifically in the Sargasso Sea. It was demonstrated at the same time that the American eel bred at nearly the same place.

Immediately after the end of the First World War, Schmidt resumed his investigations of the eel in the Atlantic. The schooner »Margrethe« had been wrecked on a reef in the West Indies in 1913 during a cruise and a considerable part of the work planned by Schmidt had therefore not been carried out. In 1920 and 1921, a very suitable ship, the motor schooner »Dana« was made available by the East Asiatic Company in Copenhagen for cruises to the Sargasso Sea in order to investigate the precise areas where the European and the American eels were assumed to breed. A very large number of eel larvae were collected including a number which were freshly hatched.

For many people it seemed that Schmidt's work had succeeded completely in demonstrating the breeding place of the eel. He, himself, however, felt that some questions were still left unsolved. For example, very convincing evidence could come from the hatching of the eggs of eels. An investigation on

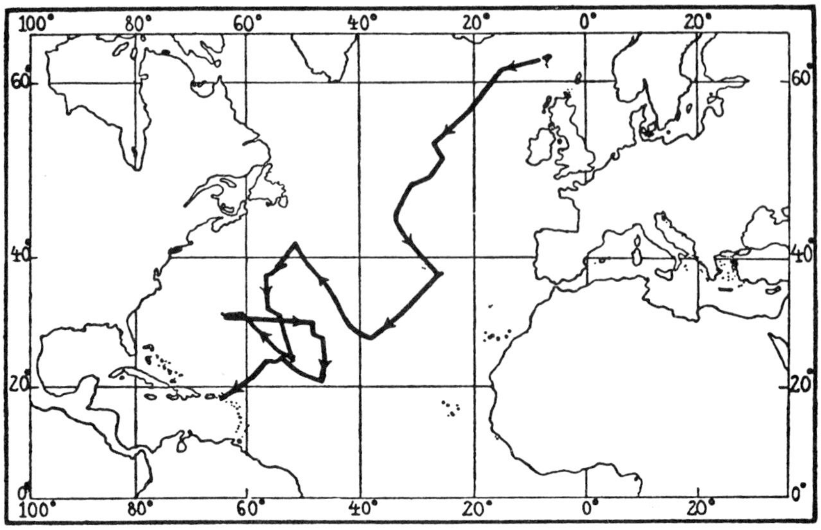

84 Route of schooner »Margrethe«, 1913.

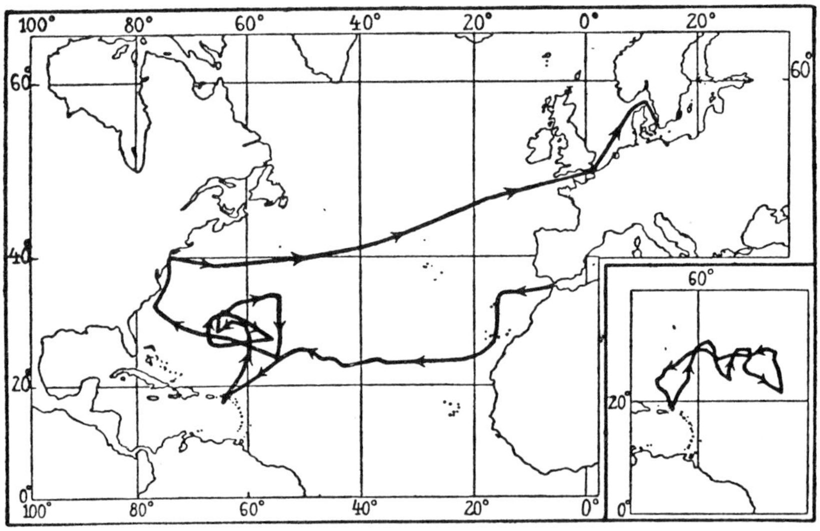

85 Route of »Dana I«, 1920. Insert: »Dana I«, 1921.

which Schmidt had set great hopes concerned the hatching of certain eggs which he suspected to be the eggs of the freshwater eel. In this, however, he did not succeed and this was a real disappointment. In view of the difficulties encountered in ex-

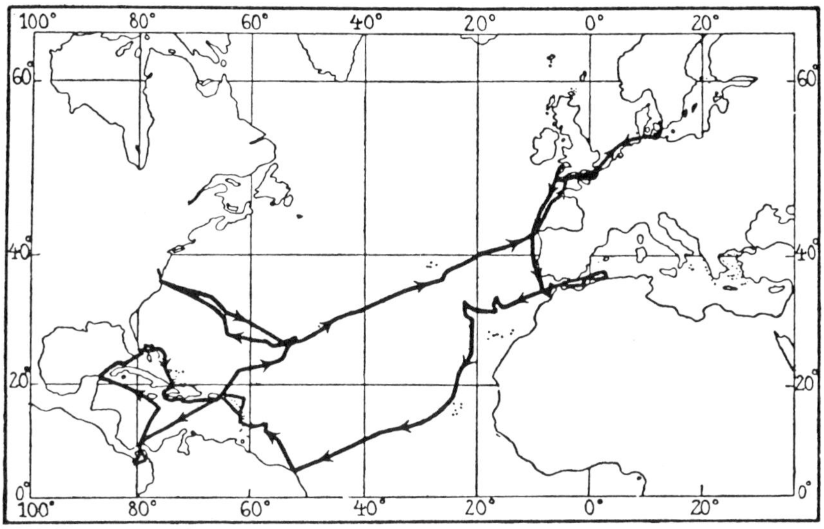

86 Route of »Dana II«, 1921–22.

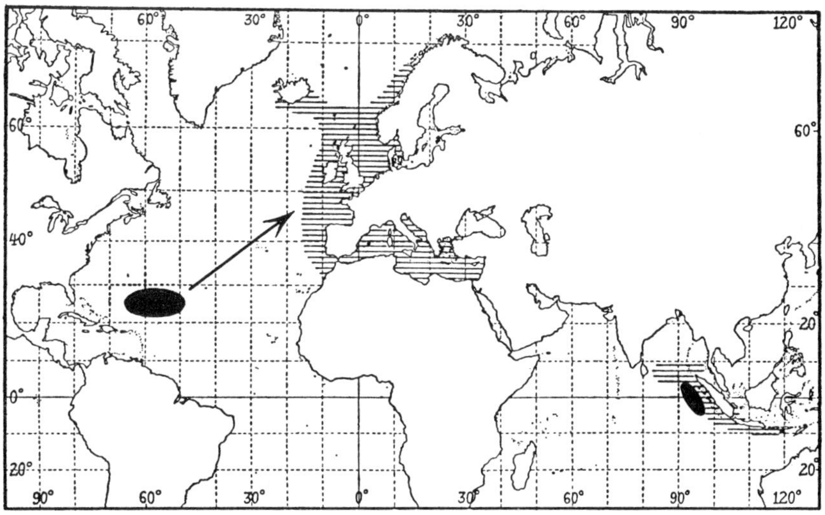

87 Breeding grounds of the European eel (Sargasso Sea) and the Sumatra eels. The striated areas denote the wandering areas of the metamorphosed elvers.

periments carried out much later with much more advanced techniques, the results were not surprising.

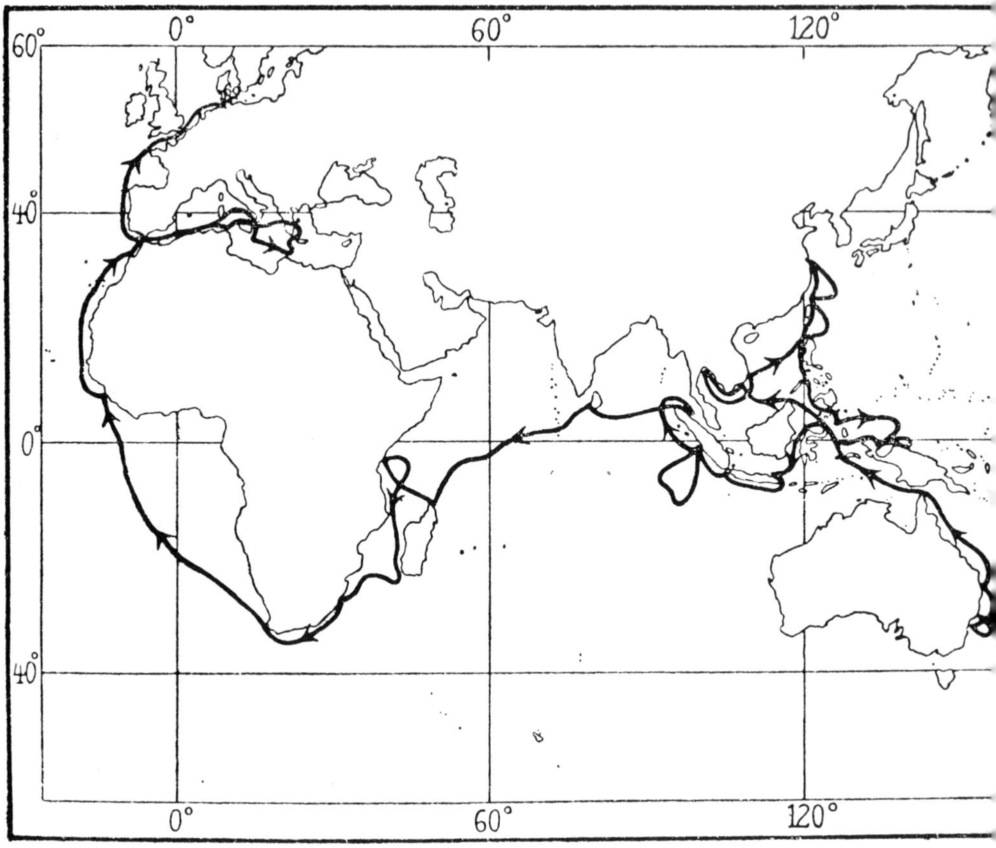

88 Route of »Dana II's« circumnavigation of the globe 1928–1930.

A new expedition was undertaken with the principal aim of investigating the general oceanographic conditions in the central part of the north Atlantic. He now had at his disposal a new ship which was especially equipped for work in the deep ocean. The Danish government had bought an English mine-sweeper which had been transformed to be used for all kinds of marine research. It was named »Dana« (II) like the motor schooner which had preceded it. This voyage, which lasted almost a year (1921–1922) was mostly financed by private individuals. The Carlsberg Foundation did, however, support an excursion of several weeks in the Pacific which was

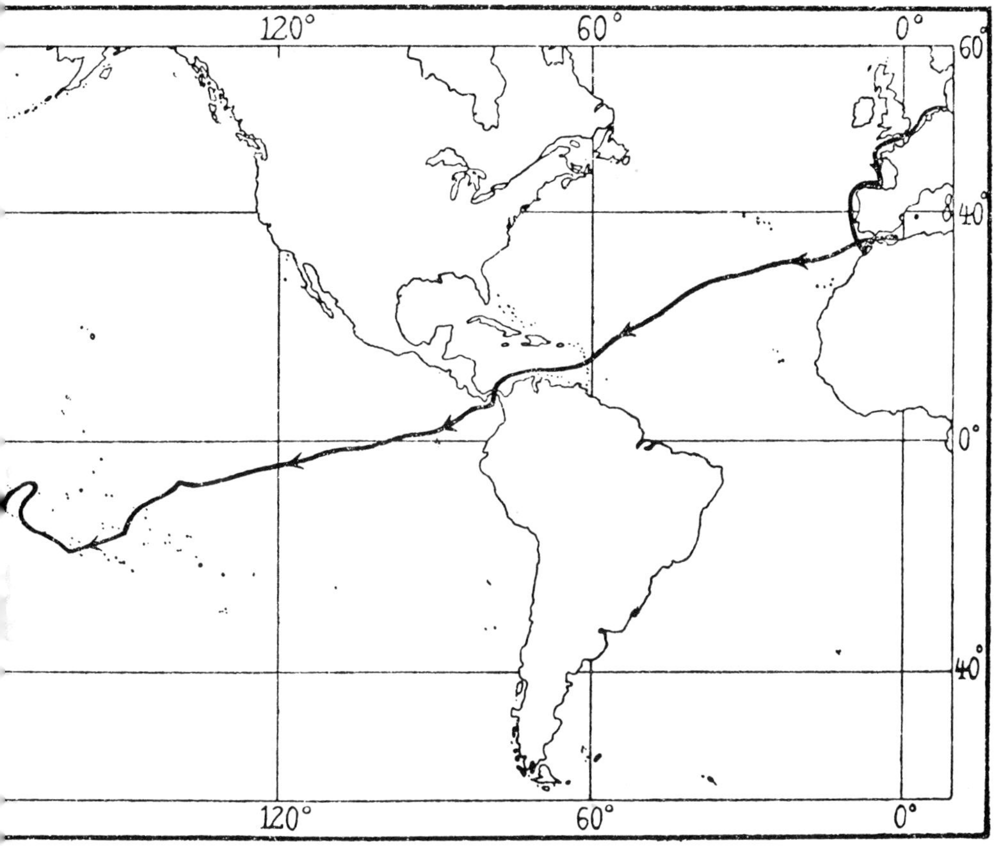

reached by way of the Panama Canal. This expedition whose members included a number of scientists from various disciplines was very successful. The cruise into the Pacific was especially interesting. The hydrographic conditions near the Gulf of Panama were found to be completely different from those in the Atlantic. At a depth of about 500 meters, a water layer was found which was practically devoid of oxygen. During the next »Dana« expedition (1928–30) this water layer was followed far out into the Pacific.

The large collections of all kinds of animals found in the pelagic water masses of the Atlantic made it clear that the classification and analysis of this huge mass of material would establish Denmark as the leading nation in this branch of oceanography. Schmidt was, however, not content with concentrating on the Atlantic. The small cruise into the Pacific had told him that similar collections from other oceans would give the work a much wider background. Therefore, he started to prepare another expedition covering all oceans. As before, his efforts in raising the necessary funds were successful.

The freshwater eels were, of course, not forgotten. Schmidt had managed to elucidate the biology of the two Atlantic species. But other species were found in the central and western part of the Pacific and still others in the coastal regions of the Indian Ocean. As the knowledge concerning the taxonomy of the mature eels was very fragmentary, he had first to solve this problem. The material available in various museums turned out to be far from sufficient. To make up for this, he started collecting freshwater eels from all over the world, whereever these fishes were found. In 1926, he went on a trip around the world, traveling as a passenger on commercial ships. Even in this guise, he made substantial contributions to oceanography. In Australia, New Zealand and Tahiti, he collected eels himself. In other regions he persuaded people to organize the collection.

Another round the world expedition in 1928 contributed further to the solution of this problem. It provided evidence for the completion of the taxonomy which was only finished after Schmidt's death. By this time, it was no longer the distribution of the mature freshwater eels which was the object of the investigation but the breeding places. As 15 species of freshwater eels are found in the Indo-Pacific region alone, the task was quite large.

The official name of the 1928 expedition was »The Carlsberg Foundation's Oceanographic Expedition Round the World 1928–30«. The Danish government had put »Dana« (II) at the disposal of the expedition and the Danish navy provided the officers and a part of the crew. All other expenses were paid by the Carlsberg Foundation. The Carlsberg

89 »Dana II« entering the harbour of Copenhagen after the expedition 1921–22.

Laboratory in Copenhagen was, of course, the headquarters for the expedition. All the large collections gathered by the expedition were from time to time shipped to Copenhagen and deposited in the Carlsberg Laboratory.

The »Dana« crossed the Atlantic to Panama, went through the canal, and crossed the Pacific to Sydney, Australia. From here the ship sailed to the north as far as 32°N. Afterwards, it crossed the Indian Ocean.

During this part of the voyage a discovery was made which later should prove to have far-reaching consequences. Between the Maledive and Seychelle islands the echo sounder showed that the ship passed an area with depths of less than 3000 m. Furthermore, measurements of the deep sea temperatures showed significant differences to the east and west of this rise. From this it could be concluded that the echo sounding did not indicate an isolated mountain top, but a submarine chain of mountains, running from NW to SE, and separating the bottom water for a long distance. The ridge thus found was named »Carlsberg Ridge«. Much later it turned out that this ridge was one link in the global system of submarine ridges (the Indian ocean analogue to the Midatlantic and East pacific ridges) which were of decisive importance for the modern hypotheses about the formation of the earth's »continental plates«.

The »Dana« then went round the south of Africa, returned via the eastern Atlantic and the Mediterranean, and after two years was home again in Copenhagen.

Pelagic fauna and flora were collected during the whole trip. Various fishing gear was used, operating down to depths of more than 3000 metres. Areas which were suspected of being the breeding places of the freshwater eel were intensively investigated. The weather situation in the north-eastern part of the Pacific prevented the search for the expected breeding place of the Japanese eels. It was possible, on the other hand, to make a detailed investigation of an area just west of Sumatra, the so-called Mentawei Deep, where all four Indo-Malaysian freshwater eels turned out to have their breeding place.

An expedition of the size of that made by the »Dana« in 1928–30 cost a great deal of money. The working up of the huge collections brought home from the expedition, however, cost considerably more. The sorting of the material had already started when the expedition arrived home. Every shipment had to be sorted in order to send the various groups of animals to the appropriate specialists. Unfortunately, Johannes Schmidt died rather suddenly while the work was in progress, less than three years after the return from the expedition. He had, in fact, not been in good health for several years.

The succeeding head of the Physiological Department of the Carlsberg Laboratory was not an oceanographer. Nevertheless the Carlsberg Foundation continued to support the work under the leadership of Schmidt's assistant, Dr. Å. Vedel Tåning. After a short interval during which the collections were stored

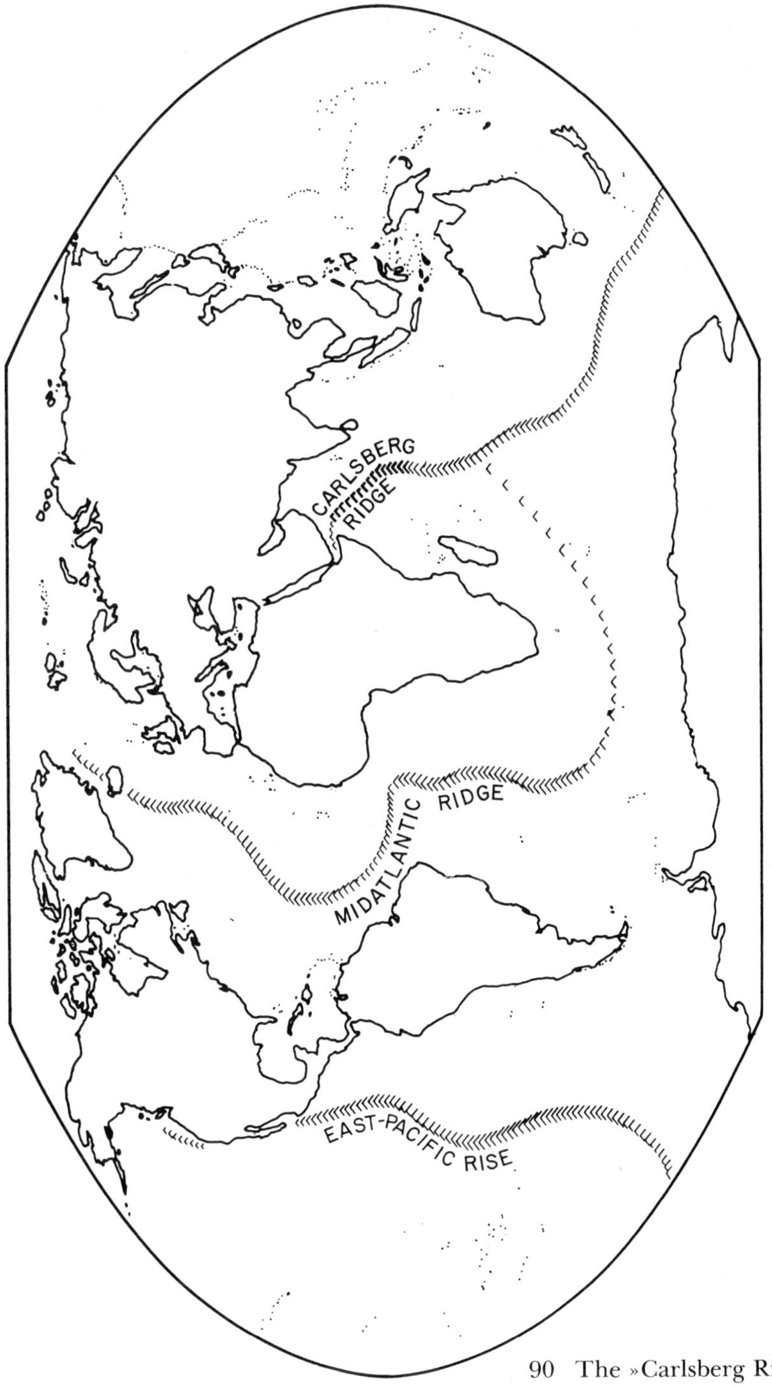

90 The »Carlsberg Ridge«.

91 Dana collections. Thousands of samples from the expedition 1921–22 in the attic of the Carlsberg Laboratory.

in temporary localities, they were transferred to »Charlottenlund Slot«, which the Danish Government had remodeled to house both the Danish Institute for Fishery and Marine Research, and the International Council for the Exploration of the Sea.

The collections from all Schmidt's expeditions were placed in large rooms in the basement, and laboratories were established where a thorough sorting of all the samples took place. The work lasted nearly 40 years. A considerable part of this large material has been analyzed by Danish scientists. However, a small country like Denmark cannot have specialists for all kinds of organisms. Scientists from all over the world have participated in the study of the various groups of animals which had been collected. All the data have been published in 83 data reports sponsored by the Carlsberg Foundation.

The material from all Schmidt's expeditions is now finally placed in the Zoological Museum in Copenhagen. It stands as a memorial to the days when the Carlsberg Laboratory was an international center in the field of oceanography.

When the Swedish oceanographer Hans Pettersson wrote his book »Mysteries of the Deep Sea«*), he dedicated it to »the memory of Johs. Schmidt, Denmark's great explorer of the deep seas«. This dedication is not only an expression of Schmidt's appreciation by his colleagues; it also appreciates the fact that Schmidt's expeditions were an important link in a famous succession of deepsea expeditions round the world. These expeditions range from »Challenger« (1872–76) over »Meteor« (1925–26) and »Dana« (1928–30) to the famous voyage of the Danish ship »Galathea« (1950–52). The last-mentioned, highly successful expedition had as its main aim the exploration of the bottom fauna of the deepest ocean trenches, and it was entirely based on the experiences gained by Johs. Schmidt and his crew. The captain of the Galathea expedition, S. Greve, and its scientific leader, A. Fr. Bruun, had both sailed with Schmidt on the »Dana«, as had several members of the »Galathea's« scientific staff, among them the writer of this chapter.

Oceanographic research has a good tradition in Denmark. There is no doubt that in the creation and maintenance of this tradition the Carlsberg Laboratory has played a very important role, and that its influence can be traced to the present day, as scientific papers based on the results of the Danish expeditions continue to appear.

*) *Title translated from the Swedish »Djuphavets Gåtor«, Albert Bonniers Förlag, Stockholm 1943.*

X

The links between the brewery and the laboratory

by *Birger Trolle*

To J. C. Jacobsen the interaction between the laboratory and the brewery was a necessary condition. If this condition was not met, neither of the two institutions could thrive. In this chapter it will be shown that very soon after the establishment of the laboratory a rich interaction arose which has lasted and kept up with the times during the century of the laboratory's existence.

As mentioned previously, the first laboratory at Gamle Carlsberg was started in 1871 in a single room of a building marked C on the map*). A few laboratory instruments were set up, a chemist, G. C. Faye was engaged for a short period, but the room was also available to J. C. Jacobsen himself or to the chemists of the brewery, if they wanted to perform an analysis or examine some problem from the brewery operations.

In May 1875 J. Kjeldahl was put in charge of the lay-out of the first real Carlsberg Laboratory, which together with the Carlsberg Foundation was formally established on September 25, 1876, and Kjeldahl became head of the Chemical Department. The laboratory was set up in the still existing building B facing the railway, and 17.000 DKr. (a large sum at that time) was spent on installations, instruments and other equipment.

The Physiological Department was established the next year, but its first head, Rasmus Pedersen, only remained for one

*) *fig. 106, last in this book.*

92 S. A. van der Aa Kühle, 1849–1906.

year and was succeeded on January 1, 1878, after one year's vacancy, by the fermentation physiologist Emil Chr. Hansen. He came from the Ny Carlsberg Brewery Laboratory (M), where a similar but less decisive development had taken place, and J. C. Jacobsen's son Carl Jacobsen had engaged Hansen in January 1877 to perform microscopic investigations of wort,

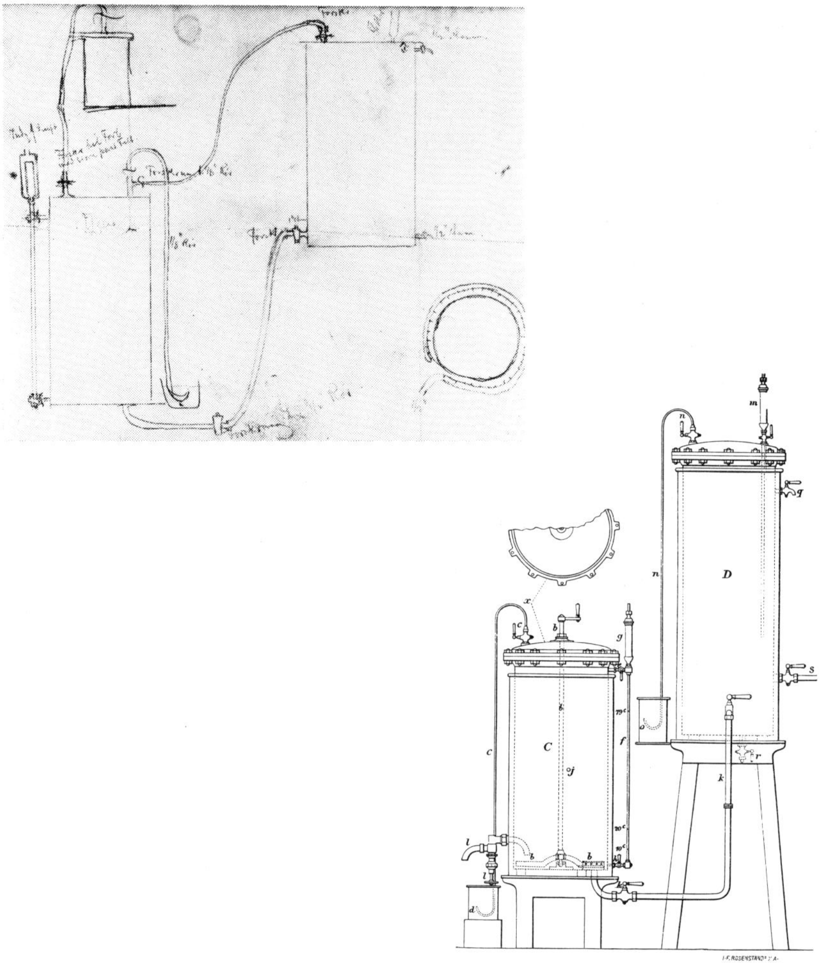

93 Kühle's hand-sketch of the yeast propagating apparatus, and a drawing showing its final shape (compare fig. 43). D: Wort container; C: Propagation cylinder; b: manual stirring device; g–h: entrance of air supply (through filter g).

beer and yeast. Hansen was succeeded later at the Ny Carlsberg Laboratory by the chemist, Lars Knudsen, and the botanist Chr. Grønlund. The pharmaceutical chemist H. Schjerning also worked there for some time, before he went to the Chemical Department of the Carlsberg Laboratory.

Thus from the very beginning there was a good and friendly contact between people in the laboratory and the leaders and chemists in both breweries, and this very soon bore fruit. Two important discoveries were published in 1883: one was the introduction of pure cultivated yeast by Emil Chr. Hansen, the other Kjeldahl's method for the determination of nitrogen in organic substances. These discoveries created a new background for the work in the brewery and thus an opportunity for intimate co-operation and at the same time world-wide fame to the Carlsberg Laboratory.

The pure yeast was introduced into brewery operations on November 12, 1883, and created absolutely safe fermentation conditions. Furnishing the brewery with rather large quantities of pure yeast each month became, however, too extensive a job for the Carlsberg Laboratory. Emil Chr. Hansen therefore approached the technical director, S. A. van der Aa Kühle, and mainly on his initiative an apparatus for continuous mass production of pure cultivated yeast was designed. It was introduced into brewing operations in November 1885 and has been in use with hardly any change ever since. As early as 1895 the apparatus was used in 157 bottom fermentation breweries all over the world.

Further, Kjeldahl's method opened up a formerly inaccessible area of analysis. It now became possible to determine nitrogen and thereby protein in barley, malt, wort and beer; a new chemical operation control had become possible and developed fast.

With these two discoveries as a background, the need for a regular scientific control of the brewery operations was increasing. In 1886 a Brewery Control Laboratory was established in a wing of a building (D) at Gamle Carlsberg, facing Gamle Carlsbergvej, and Anton Petersen was appointed head.

It was also a consequence of the co-operation between brewery and laboratory that for shorter periods both an experimental malt house and an experimental brewery were at hand. J. C. Jacobsen and E. Kogsbølle's experimental pneumatic malt house of 1800 kg capacity was in operation from 1877 to 1879, and on the basis of the results obtained pneumatic malting was introduced into Gl. Carlsberg in 1879. The experimental brewery at Gamle Carlsberg had a capacity of 4 hl and was in operation 1887–89, when it had to give way to a new brewhouse outfit.

Up to the turn of the century the bonds of interaction between laboratory and brewery had become strong, in accordance with the wishes of J. C. Jacobsen. This was partly due to the leading persons, Kjeldahl, Emil Chr. Hansen, J. C. Jacobsen and Kühle, but also to the pleasant and co-operative atmosphere at Carlsberg, which at that time was still outside Copenhagen.

When the Carlsberg Laboratory in 1897 began to work in the large new building at Gl. Carlsbergvej 10 (G), the brewery laboratory took over part of the accommodation previously used by the Carlsberg Laboratory, and simultaneously a natural co-operation also developed with the Ny Carlsberg laboratory at Ny Carlsbergvej. The latter laboratory moved in 1898 from the basement of the present Central Laboratory building (K the former main office of Ny Carlsberg) to the first floor. When Gamle and Ny Carlsberg were merged in 1906, the two brewery laboratories were united in the building at Ny Carlsbergvej (K), and hence a co-operative relationship with the Carlsberg Laboratory arose. One of the two leaders of the brewery laboratory, N. Hjelte Claussen, had come from the Carlsberg Laboratory to the Ny Carlsberg Laboratory in 1898. Hjelte Claussen, incidentally, made an important contribution to microbiology with his studies on Sarcina and Brettanomyces in the years 1903 and 1904. Also Schjerning's work on protein chemistry was of basic importance for the chemical control of the quality of beer. Most of this work was carried out during the years 1906–13 in the premises of the original Carlsberg Laboratory (B) which acted as a sub-department of the main laboratory. Schjerning did pioneer work in fractionating the

94 N. Hjelte Claussen, (1866–1955).

proteins of barley, malt, wort and beer, as well as accounting for the enzymatic tranformation of proteins during malting and brewing.

At the Carlsberg Laboratory proper, in the new building (G), work was continued in the old spirit of co-operation but still the picture changed, when Kjeldahl, whose health had been failing during later years, died in 1900. His successor, S. P. L. Sørensen opened up new channels of collaboration. He knew to guide and give ideas without anybody feeling that they were guided, and during the nearly 40 years of his leadership of the Chemical Department, he left his lasting and decisive mark upon the collaboration in the whole Carlsberg area. His influence extended to the Scandinavian School of Brewing and to the co-operation among Scandinavian brewers and brewing chemists.

S. P. L. Sørensen was working on the mechanism of enzymatic reactions. One of the most fundamental contributions to this difficult field was his paper of 1909, in which he clarified the tremendous importance of the hydrogen ion concentration (pH) for all enzymatic processes. This immediately became important for the mashing process in the brewhouse. This process had been difficult to understand, but now the introduction of pH-control gave the necessary clarity. Carlsberg was the first brewery in the world to introduce the technique of controlling pH in the brewhouse (1911–12). This was not too easy to begin with, but the difficulties were overcome by close co-operation between Sørensen and N. Hjelte Claussen, who in 1911 had returned from the United States to become technical superintendent, and later director of the breweries.

When Emil Chr. Hansen died in 1909, he was succeeded by Johannes Schmidt, who during the first years of his tenure in collaboration with his assistant Øjvind Winge worked within a new field of botanical genetics, the hybridization of hops. The idea was to obtain varieties with a higher content of the bitter principle α-acid, and within this field there was a close co-operation with the brewery laboratory with regard to chemical analyses, which unfortunately had to remain incomplete, since hops chemistry was primitive at that time. Later, when Winge succeeded Schmidt in 1933 as head of the Physiological Department, this whole question was tackled again.

Up to the early 1920s, World War I caused a stagnation in many activities, and this also influenced brewing research

95 Building of the Research Laboratory of the Carlsberg Breweries, now »Central Laboratory«.

which in many ways, especially as regards the quality evaluation of the finished beer, remained primitive and insufficient. Around the middle of the twenties brewing scientific research started abroad, aiming especially at a characterization of the properties of finished beer. This branch of research held great promise for the practice of brewing, and therefore both the Carlsberg Laboratory and the directors of the brewery considered it desirable to establish a research laboratory (K) within the brewery and in close connection with the existing control laboratory. This was done in 1928. The idea behind this new venture was a desire to keep up with and contribute to the new branch of brewing research without being bound to publish all results immediately. In compliance with a previous (1903) agreement of co-operation the »United Breweries« (Tuborg) had been informed of these plans, and consequently Tuborg in 1929 also established a research laboratory in connection with their own control laboratory.

At Carlsberg S. P. L. Sørensen acted as an informal supervisor of research, of course in close contact with the technical director, at that time N. Hjelte Claussen. The first head of the new laboratory was Niels Nielsen, but about two years later (1931) he took over the position of head of the newly established Sub-Department of Fermentation Physiology at the Carlsberg Laboratory (J). Nielsen's successor in the brewery research laboratory was Erik Helm who kept the position till 1946. Since then B. Trolle has acted as head.

The research laboratory became the basis of a very close and lasting co-operation with the Carlsberg Laboratory. At weekly meetings, also attended by the leaders of the brewery operations, S. P. L. Sørensen was in charge of the organization of the work, and many valuable contributions to new current subjects such as chemical turbidity, foam measurement, gushing, taste testing technique, besides many experiments in brewery operations, were conceived at these meetings.

When Øjvind Winge was appointed Head of the Physiological Department in 1933, he also was most positive in extending co-operation. First the brewery side kept in touch with his and O. Laustsen's pioneer work on the sexual propagation of yeast, and co-operation further increased, when Laustsen in 1939 went to the brewery laboratory. Especially important for the co-operation with Winge, however, were developments in barley breeding. For years the leading institution in this field in Denmark had been Abed Experimental Station. The first experimentally crossed barley variety, Abed Opal, appeared in 1922, followed later by Kenia and Maja. From 1933 these two barley varieties headed the Danish market for about 15 years, and the Danish barley harvest showed an annual increase of about 100.000 t owing to the greater yield of these varieties. When Winge entered the barley breeding field, the brewery bought in 1938 an experimental farm »Nordgaarden« which was administered by Winge for more than 20 years. Here new

96 Harvest of hops at the old experimental farm »Nordgaarden«.

varieties were discovered and propagated and their brewing characteristics evaluated in the brewery research laboratory. This interesting and positive co-operation resulted in barley varieties such as Lenta, Carlsberg and Nordal which dominated the market for years. In this connection it is worth recalling that J. C. Jacobsen was interested as early as 1871 in the question of the quality of barley for malting. In an application to the Royal Danish Agricultural Society he suggested the appointment of a malting barley committee. This was formed in 1883.

Further, the various Carlsberg hop hybrids which Winge had produced himself, many of which had an extraordinarily high content of α-acid, had ideal experimental and growing conditions at the experimental farm. They were tested every year in the research laboratory's experimental brewery as well as in brewery practice. These testings remained, however, unsatisfactory for many years due to the primitive state of hop chemistry. Not till the late 1950s was a satisfactory method developed for determinating the bitter tasting hop components of wort and beer.

Returning to a chronological account, S. P. L. Sørensen was succeeded in 1938 by his brilliant assistant K. Linderstrøm-Lang who together with the technical directors Hans Fogh and K. Georg Sørensen continued to supervise the co-operation between the Carlsberg Laboratory and the brewery. The problem of the chemical stability of beer greatly interested Linderstrøm-Lang, as these problems were in focus at that time. He followed the research work in the brewery laboratory, which led to the possibility of keeping the chemical stability of the beer under control. Linderstrøm-Lang supported most positively the plans for an experimental brewery (capacity 100 bottles of finished beer) which was established in 1943. He also endorsed the acquisition of a Svedberg ultracentrifuge, mounted in a special extension of the research laboratory, inaugurated in 1951 and operated by Robert Djurtoft. All these activities were aided by Linderstrøm-Lang's rare personal charm and diplomatic skill.

At the end of the 1950s Winge and Linderstrøm-Lang were succeeded by H. Holter and M. Ottesen, who had been active at the Carlsberg Laboratory for many years. They continued the co-operation with the research laboratory, since 1956 part of the combined Control- and Research Laboratory organisation and now in the hands of Poul Gjertsen, who had contributed notably to the solution of the severe gushing problems caused by the wet barley harvest in 1954. The subjects discussed at the weekly meetings now extended to include the aroma substances of beer and also the amino acids. Owing to the highly developed chromatographic analysis technique both groups of substances could be determined individually and with great accuracy. Professor Ottesen is a specialist in this area, so that the co-operation is most profitable. Enzymatic processes, the content of protein fractions of wort and beer as

97 The present experimental farm »Allindemaglegaard«.

well as the carbohydrate glucane have also been ardently discussed. Further, the occasional failure of yeast to be stable with regard to flocculation, or the character of non-flocculating yeast, has been considered, but this problem is open to future research

The subject of barley breeding is also on the future research programme, and many interesting new lines in the recent progress of brewing chemistry and enzyme chemistry are to be pursued. This future research programme is being taken care of by Professor D. von Wettstein, the coming head of the Physiological Department, in co-operation with the research director of the Department of Biotechnology of the new Research Center, Dr. Lars Munck. Thanks to the initiative of Director A.W. Nielsen, the former experimental farm »Nordgaarden«, the soil of which was not ideal, has been exchanged for »Allindemaglegaard« near Ringsted in 1965.

When Carlsberg and Tuborg merged in 1970, also the

whole laboratory organisation became changed. Since 1973 the former combined Control- and Research Laboratories at Carlsberg have been reorganized under the name The United Breweries' Central-Laboratory, now responsible for the correct quality of all Carlsberg and Tuborg beer produced both inside and outside Denmark. All research within the Company has been centralised in the new Carlsberg Research Center as reported in another chapter og this book. A close contact between the Central-Laboratory and the Research Center is foreseen as a natural continuation of the existing excellent collaboration during the last century.

In this chapter an attempt has been made to give a glimpse of the interaction between the laboratory and the brewery, and to show that this interaction in its way has contributed to fulfil J. C. Jacobsen's intentions, as expressed in his frequently quoted »Golden Words«:

»In working the Brewery Gl. Carlsberg it shall be the constant aim, regardless of immediate advantage, to develop production to the greatest possible degree of perfection so that this brewery and its product may always stand as a model and ensure by its example that the manufacture of beer in this country is maintained at a high and honourable level.«

XI
The transition from the Carlsberg Laboratory to the Carlsberg Research Center

by *P. Brandt Rehberg*

The most important event in the recent history of the Carlsberg Laboratory was its separation from the Carlsberg Foundation and its transfer to the Breweries. It was not an easy step to take and it would not have been taken without cogent reasons. In order to provide a clear picture of the reasons, the protracted negotiations, and the outcome, the editor has asked Professor Rehberg, the chairman of the Foundation during the crucial years, to recount what actually happened.

Legalistic matters and documentary evidence normally do not make entertaining reading. However, in the present case the dryness of the subject will undoubtedly be counterbalanced by its deep interest for all friends of the laboratory.

At the end of the 1960s, the economic consequences of the growing inflation forced the Carlsberg Foundation to an appraisal of its total activities. According to its statues, the Foundation should out of its yearly budget make available certain amounts of money for the two departments of the Carlsberg Laboratory (Division A of the Foundation), and likewise for the Museum of National History at Frederiksborg Palace (Division C), while Division B should dispose over the remaining sums which should be used for the furthering of science in general within certain limitations laid down in the statutes.

In addition to the original Carlsberg Laboratory's two departments, the Foundation (assisted by a grant from the Rockefeller Foundation for part of the building expenses) established in 1931–32 a new laboratory, the Biological Institute of the Carlsberg Foundation. The funds for the running expenses of this institute, situated in Tagensvej in Copenhagen, are taken out of the budget for Division B of the Foundation which over a long period had enough money for the running of this laboratory in addition to its main commitments.

In the late sixties, however, the running expenses of these three laboratories (the two Departments of the Carlsberg Laboratory and the Biological Institute) began to cause trouble for the Foundation. If the Carlsberg Foundation runs a laboratory, it must be an absolute condition that such a laboratory in all respects shall be fully up to the standard of similar laboratories, for instance the scientific laboratories of the State Universities. The level of the salaries paid at the Carlsberg Laboratories is laid down in the original statutes: »The head of the laboratories shall be given a stipend at least on a par with that of a University professor ...«, and this equality with comparable stipends must of course hold good for the rest of the personnel as well. It has therefore been normal practice that the salaries at the Foundation institutes are adjusted according to the state salaries for similar institutions. In later years the state salaries have increased considerably, so that the Foundation has been forced to keep up with this increase. At the same time, research in the natural sciences has undergone a rapid technical development in which, among other things, automatic electronic equipment plays a greater and greater role. These instruments have improved the analysis of biological samples to such an extent that laboratories not furnished with

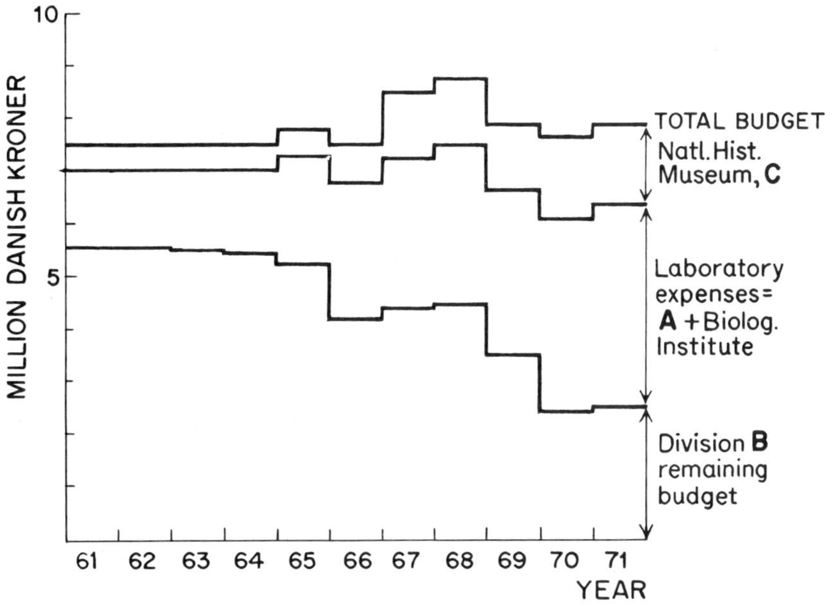

98 Expenditures of the Carlsberg Foundation during the years 1961–1971.

such equipment will fall hopelessly behind. The Carlsberg Foundation has therefore been forced to participate in the scientific equipment race, which unfortunately is a costly affair. Not only the procurement expenses but also the cost of maintenance is high, while the life time of the modern equipment is often shorter than that of the simpler instruments of older times.

These combined circumstances have forced the Foundation to use a steadily increasing part of the available money for laboratory expenses. As shown by the accompanying graph, it could be predicted that the amount of money available for the purposes of Division B would decrease every year and would in the course of some few years become so small that the balance between the expenses for the laboratories and for the other activities of the Foundation would be seriously upset.

It is of course impossible to compare the value of the different activities of the Foundation, but anyone familiar with the history of Danish science in the last century will know the enormous importance of the assistance given through the scientific grants of Division B.

The Board of the Foundation was bound to consider it of the utmost importance that this part of the activities of the Foundation was not seriously weakened, and it was therefore necessary to find ways of reducing laboratory expenses, without diminishing the quality of the work done. The only possible course seemed to be a reduction in the extent of laboratory activities.

The formal relations between the Foundation and the three individual laboratories were rather dissimilar. While the two departments of the Carlsberg Laboratory had their finances and management firmly laid down in the statutes of the Foundation, the Biological Institute was established by the Board of the Foundation, and the rules for its finances and management were laid down in separate statutes issued by the Board and subsequently approved by the Academy of Sciences and Letters. It seemed natural in the first place to consider the possibility of discontinuing the Biological Institute, the more so as it seemed feasible to transfer all the activities of this institute with its entire staff to the Physiological Department of the Carlsberg Laboratory. The background for this possibility was that the time for retirement of the head of the Physiological Department, professor Heinz Holter, was approaching, and that the head of the Biological Institute, professor Erik Zeuthen, seemed to be Holter's natural successor because of his whole training and field of work.

The possibility for a solution along these lines was carefully examined. The plan implied that the building of the Biological Institute with its site should be transferred to the University of Copenhagen, accompanied by financial compensation, which would be most welcome as a contribution towards the much needed modernization of the Carlsberg Laboratory.

In March 1971 the scientific personnel of the three laboratories involved were informed of the difficulties foreseen by the Foundation, if it should continue to operate all three laboratories. The possibility of a transfer of the Biological Institute to the Carlsberg Laboratory was mentioned, but the discussion revealed certain difficulties, primarily due to the commendable fact that the personnel of all three laboratories felt themselves intimately bound to the spirit and traditions of their own laboratory. In particular the staff of the Biological Institute were uneasy about giving up their status as quite independent researchers and accepting the obligations associated with working in the Carlsberg Laboratory.

Moreover, it turned out that the ownership of the site of the Biological Institute was linked to the operation of a biological laboratory by the Foundation, and that the site, if not so used, had to be returned to the Ministry of Defence from which it had originally been procured. In that case the Foundation would even have had to pay for the removal of the building. The transfer of the building to the University of Copenhagen thus had to be given up, and because of this and other difficulties the whole scheme was abandoned. Another plan, which implied the tranfer of the Physiological Department of the Carlsberg Laboratory to the Biological Institute, proved impossible from the beginning, as it would have been in conflict with the whole idea of the Carlsberg Laboratory as conceived by J. C. Jacobsen. Article 25 of the statutes clearly states »that this institute may never be separated from the property »Carlsberg«, as long as a brewery is operated there. Neither may the laboratory ever be amalgamated with another institute«.

At the same meeting in March 1971, the head of the Chemical Department of the Carlsberg Laboratory, professor Martin Ottesen, had suggested the possibility that the Carlsberg Breweries should take charge of the Carlsberg Laboratory and of the operation of this institute, a solution which he had discussed with the managing director of the Breweries, Mr. A. W. Nielsen and with others. This solution, which had been previously and tentatively discussed within the Foundation board, was now seriously considered.

Financially, it would mean that the cost of running the laboratory, now beeing a part of the deductible brewery expenses, would make itself felt with much less weight in the net returns flowing from the brewery to the Foundation; the total result would therefore be a significant saving in the resources available for the purposes of the Carlsberg Foundation.

Most important for the Foundation was of course the question whether a transfer to the Carlsberg Breweries, recently amalgamated with the Tuborg Breweries (under the name »United Breweries«), would be compatible with securing for the scientists at the laboratories that scientific freedom which through all the years had been the foundation of the high scientific standard of the laboratories and of the high esteem they have always commanded. It was moreover necessary once more to scrutinize carefully the intentions of J. C. Jacobsen in establishing the laboratory. What was intended to be its field of work, and what would be the conditions for its work in the new situation?

The old brewer proved to be, in this field also, far ahead of his contemporaries in his understanding of the conditions necessary for a fruitful cooperation between science and industry. As early as 1871, J. C. Jacobsen had established a service laboratory within the Carlsberg Breweries, but he was not content with this. He understood that a laboratory of this kind would have a far too restricted field of work, and that because of this its staff would not get the opportunity to develop, perhaps even hardly to preserve, their scientific abilities. Scientists should not be encumbered with routine work, nor with elementary teaching, both of which would leave no time for true scientific research. It was ideas of this kind that led to the foundation of the Carlsberg Laboratory. The brewer's intention in doing so was quite clearly to create a free, broadly based research institute with which the breweries could cooperate, so that it might in the best possible way fulfil its tasks, among which were: To contribute towards a better understanding of the processes of brewing.

It is amazing to find as early as in 1876 an understanding so conscious of this point of view, for which we have to fight even to-day, namely that the basis for all industrial or other applied research is free basic research. For Jacobsen it was »a matter of course« that scientists working in such a specialized institute »in addition to their work on the special subject of the institute, should strive to maintain their scientific ability in general by other studies and investigations, but in such a way that the main task of the institute is not neglected or forgotten«. It was indeed a free but specialized research institute which Jacobsen intended to create, and as such it has been operated through all its existence.

The problem was now whether it would be possible to transfer the laboratory to the United Breweries under such conditions that this status of the laboratory was upheld, and in such a way that all persons and institutions involved were satisfied.

In the first place, would the United Breweries be interested in taking over the laboratory and its future operation with all the strings necessarily attached to the arrangement? It was not enough that the Carlsberg Foundation agreed and that the representatives of the Foundation on the Board of the Breweries could vote for the plan. Though the Foundation holds a majority in the Brewery Board, it was absolutely necessary that the minority representatives, representing the original stockholders, were convinced that it was in the interest of the Breweries to take over the laboratory, and that they could therefore accept the plan. It was also important in this connection that the Breweries, if the transfer could not be accomplished, were bound to engage in a large expansion of their own existing research facilities. As a matter of fact, the Managing Director of the Breweries, Mr. A. W. Nielsen, had for some time considered such an expansion, comprising both fundamental and applied research, very much along the lines laid down in the statutes of the Carlsberg Foundation or even surpassing them. Mr. A. W. Nielsen showed a complete understanding of the vital necessity for free fundamental research within such an institution, and he recognized that this alone could guarantee that highly qualified scientists would serve in the future laboratories. His thoughts in this direction were no doubt highly influenced by his talks with various scientists, notably the well-known geneticist, the late Mogens Westergaard. The chances for a transfer thus seemed promising.

Though the Board of the Breweries was informed of these discussions, it was not possible to open formal negotiations until the plan had been accepted in principle by the two authorities through which every change in the statutes of the Foundation has to pass: The Royal Academy of Sciences and Letters and the Ministry of Justice*.

Would it be possible to include in the transfer documents clauses of security for the future of the Carlsberg Laboratory, for instance as regards the right to free fundamental research, so that The Royal Academy would feel safe in approving the transfer and the necessary changes in the statutes, and would the Ministry of Justice »confirm« these changes? Even the Board of the Museum of National History (Frederiksborg Museum) had to be asked but was not expected to oppose the changes, as these could not be expected to affect the interests of the museum, at least not adversely.

First the Ministry of Justice was approached and could in principle accept the plan of transfer and the necessary changes in the »royally confirmed« statutes. Subsequently the Foundation laid the proposal before the Academy at a meeting on May 14, 1971. Simultaneously, the minority representatives of the Brewery Board were informed of the negotiations proceeding.

* The Ministry of Justice must be asked to approve any change in the statutes, because these are »royally confirmed«, a formality which gives them a high legal status but at the same time makes changes difficult.

The Academy unanimously gave its consent to the intentions of the Foundation, and it was thus possible to open formal negotiations with the breweries in order to reach a binding agreement. In this agreement the United Breweries would consent to take over the Carlsberg Laboratory and the responsibility for its future operation, and the Breweries would, as was most important for the Foundation, assume the same duties which had hitherto been borne by the Foundation, namely to respect the right of the scientists to a certain measure of free and unrestricted research, and the obligation stated in Jacobsen's rules not only as a right but as a duty: To publish all scientific results.

At the beginning of the negotiations it was intended simply to remove all those parts of the statutes of the Foundation dealing with laboratory matters and to include them in the agreement with the Breweries with the slight amendments necessary.

The Ministry of Justice, however, did not accept the removal of these articles from the Foundation statutes, and it was therefore necessary to reopen deliberations with the Breweries. During these negotiations a doubtless better solution was reached, which implied that the articles dealing with laboratory affairs should be included in identical form both in the Foundation statutes and in the agreement between the Breweries and the Foundation. This solution means that the Foundation not only has its share in the responsibility for the Carlsberg Laboratory in its capacity of shareholder in the United Breweries. In addition, the Foundation itself assumes a special right and obligation as guardian of the laboratory rights. When this solution had been accepted by the management of the Breweries, a complete draft of an agreement and of the necessary statutory changes was formulated. This draft was accepted by both parties. The substantive clause of the statutes dealing with the laboratory was formulated identically for the Foundation and the Breweries, and runs: »to continue and expand the activities in the Chemical and Physiological Carlsberg Laboratory, founded by J. C. Jacobsen in 1875 and transferred to the United Breweries in 1972, in agreement with the statutes of this laboratory, to control the management and activities of the laboratory and to contribute financially to these activities«.

The thus established financial contribution from the Foundation to the laboratory activities was fixed at a sum equal to 10% of the yearly budget of the Foundation, and this contribution was to be used for laboratory activities in fundamental research. The money, therefore, was primarily intended for the support of activities in fundamental research which due to their nature were so remote from industrial research that financing by the Breweries would be difficult to justify; but on the other hand, the money may also be used for fundamental research in general.

The proceedings were now so far advanced that the matter could be formally laid before the Academy of Sciences and Letters which at a meeting on March 24, 1972 gave its final approval. After the consent of the Board of the Museum of National History, the Board of the Carlsberg Foundation and the Board of the United Breweries were able to sign the final agreement on April 25, 1972. The following statement occurs in the preamble of this document*.

»Recognizing
the importance which an expansion of the collaboration between the »Carlsberg Laboratory« and the research establishments of the Company will have for the Company, the obligation already assumed for active coperation with the »Carlsberg Laboratory«
that an integration of the Carlsberg Laboratory within the Company will create a broader foundation for a continuation and expansion of the activity of the laboratory to the benefit of

* translation by the author of this chapter

both the Company and the laboratory, the contracting parties have agreed as follows:
§ 1. The Foundation transfers, on the conditions mentioned below, the Foundation-owned Chemical and Physiological Carlsberg Laboratory to the Company. The Company commits itself to allow the laboratory to be managed and run in agreement with the rules laid down in the statutes of the Foundation.«

The transfer included the site Gl. Carlsbergvej 10 with the laboratory buildings, the newly built laboratory for isotope research, the library, and the equipment for the Chemical Department. For these effects the company paid 6.75 mill. DKr, while the main part of the equipment from the Physiological Department, which to a large extent had been procured through American grants to professor Holter, was transferred to the Biological Institute together with professor Holter and his coworkers, who for the future would be associated with this laboratory.

The United Breweries moreover took over the obligation to continue the publication of »Comptes Rendus des Travaux du Laboratoire Carlsberg«.

Seen from the point of view of the Carlsberg Foundation, this agreement means the removal of a considerable financial burden. It has been shown earlier in this chapter how great a part of the Foundation's resources was monopolized by the steadily mounting costs of running the laboratories. By far the greater part of this financial obligation has since 1972 been taken over by the United Breweries.

On the other hand, the transfer of the Carlsberg Laboratory to the Breweries means of course that the ownership of the laboratory has passed from the Foundation to an industrial enterprise, the Unites Breweries Ltd. But in this connection it should not be forgotten that the majority of the shares of United Breweries Ltd. is in the hands of the Carlsberg Foundation. So for this reason alone there is every assurance that the voice of the Foundation will be heard if, in spite of the safeguards embodied in the agreement, questions concerning the fundamental rights of the laboratory should ever arise. This is also clearly indicated by the composition of the Board of Trustees responsible for the management of the laboratory. It is part of the agreement between the Breweries and the Foundation that this board shall consist, as hitherto, of three members of the Board of the Carlsberg Foundation, supplemented with two adjunct members elected by the Academy of Sciences and Letters. The only innovation consists in the fact that the adjunct members are now proposed for election by the management of the Breweries.

Simultaneously with taking over the Carlsberg Laboratory, the United Breweries have considerably increased their own research and development activities. This resulted in the creation of an entirely new research organization, the Carlsberg Research Center. The important role of the Carlsberg Laboratory in this Research Center is discussed in the next chapter of this book.

XII
The Carlsberg Laboratory 1976

by *Christian Crone*

It has been stated in the preface that in this book an attempt would be made to restrict technical language in the interests of general readability. In some parts of the following chapter, however, this »promise« will not be kept.

In contrast to all the other chapters, professor Crone's contribution is not primarily concerned with history. Instead, it is a report on the present state of the laboratory and even to some degree an account of plans for the future. It was therefore felt that in order to make such an account both correct and worth-while it was inevitable to go into some detail and to use the appropriate technical language.

However, this does not apply to the whole chapter. It is therefore hoped that those readers who are not especially familiar with biochemistry, plant physiology and genetics, will still find interest in the sections dealing with the present overall status of the laboratory.

The preceding chapters describe the history of a scientifically successful biochemical-physiological research laboratory. Important discoveries have been made and valuable concepts have been formulated. Centered around devoted and talented leaders, a collective effort has placed the Carlsberg Laboratory in the first rank of research institutes. The first 100 years have been surprisingly calm and, looking back, one gains the impression that the good spirit in which the laboratory was created has prevailed throughout. Many scientists outside the relatively small Danish research community have witnessed the laboratory's progress, and have personally joined in the development with their own scientific contributions. A jewel has been cut, but, unlike a jewel, its continued existence depends on a steady input of work, imagination, and up-to-date facilities. The recent expansion of laboratory space and the acquisition of new and modern research equipment, together with a significantly augmented staff, allows the centenary to take place in an atmosphere of great expectations for the future.

It is my privilege – on behalf of the Board of Trustees of the Carlsberg Laboratory – to sketch the present and thus to look into the immediate future. This cannot be a vista covering any substantial length of time, since by the very nature of science nobody can make safe projections into any distant future.

The laboratory to-day consists, as before, of the two original departments of chemistry and physiology. But around this original laboratory are now grouped others. During the last years, the scientific activity of the brewery (the United Breweries Ltd.) has been extended with three new sections for applied research, the *Carlsberg Research Laboratory,* in which brewery research in the widest sense takes place. These new laboratories, together with the two departments of the original Carlsberg Laboratory, constitute the *Carlsberg Research Center.*

Since this book describes the history of the Carlsberg Laboratory itself, special emphasis will be placed on the present status of the research in the departments of chemistry and physiology, where basic research is pursued.

The Department of Chemistry

The department is headed by Professor *Martin Ottesen,* who is now 55 years old. Ottesen's research activity has always been centered around the chemistry of proteins. Since a comprehensive evaluation of his earlier activities has been given in chapter 4, only his recent accomplishments and research interests are mentioned here.

His interest in the conformation of proteins in solution has continued, and recent studies deal with the deuterium hydrogen exchange of proteolytic enzymes from bacteria. Both the particularly alkali-resistant subtilisin type Carlsberg, which he discovered in 1951, and the related subtilisin type Novo exchange the hydrogen atoms of their inner cores more slowly than do the hydrogens of other proteins, indicating that these two proteins have very compact structures. The hydrogen exchange technique has recently been further refined to permit measurements of the small differences between the exchange rates of hemoglobin with and without oxygen. In collaboration with *Kirsten Biedermann* the nature of the proteolytic cleavage process is being studied with a model peptide in the form of ß-lactoglobulin which has been unfolded by reduction and carboxymethylation under mild conditions. Cleavage experiments with two enzymes having widely different specificities, trypsin and bromelain, indicate that they both initially attack the carboxy-terminal part of the peptide chain. This behaviour can be explained if it is assumed that the carboxy-terminal part of the chain has a greater degree of flexibility than other parts of the chain and thus has a greater chance of rearranging itself in the specific patterns which are required to make effective contact with the catalytic sites of the enzymes.

Insolubilization of enzymes by chemical coupling to insoluble matrix substances frequently leads to significant variations of their kinetic parameters. *Birte Svensson* has studied the influence of matrix substances with positive and negative net charges on the enzymatic properties of the coupled subtilisins, and the investigations have been extended to include small, water-soluble molecules which chemically are composed of the same building-blocks as the high molecular, insoluble matrix-substances. The results indicated that the major changes in the enzymatic properties of the subtilisins were not caused by the chemical reactions used for the couplings, but by the actual change from soluble to insoluble forms. More detailed characterization of the properties of insolubilized enzymes has been complicated by the fact that the individual enzyme molecules are presumably bound through different side-chains to the matrix, so that the catalytic centers of the enzymes to varying degrees are sterically blocked and prevented from coming into contact with substrate molecules. However, by using crystals of enzymes and subsequently immobilizing the enzyme molecules within the crystal structures by means of chemical cross-links, it becomes possible to produce insolubilized enzymes where all the molecules, except those in the surface region, have the same surroundings and thus the same properties. Subtilisin type Carlsberg, crystallized from sodium sulphate, has now been insolubilized by *E. Tüchsen* by cross-linking with glutaraldehyde. The kinetic properties of these enzyme crystals differ markedly from those of the enzyme in solution and, furthermore, there are indications that at high substrate concentrations, the enzyme in the crystal phase changes to a conformation with different properties. *S. Bayne,* a guest from the University of St. Andrews, Scotland, is making similar investigations of the more complicated enzyme, lactate dehydrogenase, which consists of four sub-units. Using a different cross-linking procedure, he has obtained insolubilized enzyme crystals which retain a large fraction of the enzymatic activity.

Ottesen's collaborators for several years, *Ib Svendsen* and *Jack Johansen,* are both engaged in fundamental research of the structure and function of enzymes. The research interests of Ib Svendsen focus on chemical modifications of the subtilisins. Although these proteases are less efficient than mammalian proteases when acting on small substrate molecules, they approach or surpass the efficiency of the mammalian proteases when acting on long peptide chains. This indicates that the »secondary binding sites«, i.e., binding sites at a relatively long distance from the peptide bonds being cleaved by the enzymes, are relatively more important for the function of the subtilisins than for the function of the mammalian enzymes. For this reason, chemical modifications of the amino acid side chains which are parts of the secondary binding sites have a pronounced influence on the enzymatic properties of the subtilisins. Ib Svendsen has especially studied the properties of nitrated subtilisin and demonstrated that the nitration of a single residue, now identified as a tyrosine residue in the distant sub-site S_4, was sufficient to increase the rate of splitting of the polypeptide substrate clupein several fold. Recent experiments indicate that it will be possible to identify specific carboxyl and hydroxyl groups located in different sub-sites, which are of importance for the proper functioning of the subtilisins.

The principle of isopycnic density determinations in density gradients introduced by K. Linderstrøm-Lang in 1937 was modified in the 1950's by Professor J. Vinograd and coworkers, who used stable gradients formed in strong centrifugal fields to separate large nucleic acid molecules. Professor *J. B. Ifft,* University of Redlands, California has used this technique to determine the buoyant density of protein molecules, and by titration experiments he established correlations between the densities and the net charges of the proteins. During his visit to the department in 1969/70 he and Ib Svendsen investigated how chemical modifications, which varied the net charge of proteins, would influence their densities. This collaboration has been continued during Ib Svendsen's one year visit to the University of Redlands. After his return he has demonstrated, in collaboration with M. Ottesen and *T. Graves Pedersen* that the method can be applied to cross-linked protein crystals with high precision and can provide information about both homogeneity and degree of cross-linking. The behaviour of protein crystals in density gradients will be the subject of further studies when Professor Ifft returns in September 1976 for his second sabbatical visit to the laboratory.

Jack Johansen is mainly interested in the enzymes carbonic anhydrase and carboxypeptidase with the object of clarifying important steps in the catalytic processes. The studies on carboxypeptidase have partly been carried out in Professor *B. L. Vallee's* laboratory at Harvard Medical School. »Reporter«-groups coupled to specific, single amino acid side chains in the enzyme molecule have been used to clarify the structural difference between the enzyme in solution and in the crystal state. It is now clear that the enzyme undergoes a considerable conformational change upon dissolution, a finding which has an important bearing on deductions about enzymatic processes from the structure of protein crystals. Some of the modified enzyme derivatives are chromophores and have proved very effective as spectrokinetic probes of catalytic events. These properties have been utilized in a joint collaboration with Professor Vallee and Dr. *A. A. Klyosov,* Moscow State University, to map the topography of the binding sites for inhibitors and substrates in carboxypeptidase A. Furthermore, application of rapid flow techniques has shown that such chromophores are particularly informative in monitoring multiple substructures of enzymes in solution. In collaboration with Dr. *D. S. Auld,* Harvard Medical School, the detailed kinetic properties of carboxypeptidase Y from yeast are being investigated using fluorescent-labelled peptide and ester substrates in conjunction with rapid transient kinetic techniques. This approach allows examination of kinetic events at the moment of catalysis.

A collaborative project with *R. Bauer* and *P. Limkilde,* Niels Bohr Institute of Physics, aims at using an advanced physical technique for characterizing the active site of carboxypeptidase. By means of »Perturbed angular correlation technique«, differences between the enzyme in solution and in the crystal state have been found. The same technique has been applied in studies on carbonic anhydrase where the nuclear quadrupole interaction between ^{111}Cd, bound to the active site of the enzyme, and the immediate surroundings has been determined at various pH's. The measurements have shown that a reversible change of structure takes place in the neighbourhood of the Cd-atom, involving of a group with a pK of 9.5. In the native Zn-containing enzyme, however, the pK-value is 7.3. The results of titration experiments support a Zn-H_2O to Zn-OH mechanism, where Zn-OH is the enzymatically active structure. Furthermore, these studies, in conjunction with kinetic investigations, have given new information regarding the structure and chemical nature of the active site essential for the understanding of the catalytic mechanism.

In connection with these investigations new affinity gels with high binding capacities were synthesized and used in preparing the carbonic anhydrase isoenzymes from human red blood cells and carboxypeptidase Y from baker's yeast. The last mentioned of these enzymes is a useful tool for amino acid sequence determinations and *B. Martin,* Ontario, Canada, in collaboration with other members of the department has further developed this type of method. *F. Poulsen* has used the principle of »affinity labelling« to investigate interactions between both enzymes and substrates and antibodies and antigens. The separation methods which have been used in the detailed characterization of the various stages of the proteolytic degradation process have been applied by *S. Bech Sørensen,* in combination with immunoelectrophoresis, to study the high molecular substances found in beer. *H. Abrash,* California, USA is currently investigating the pathway of the alkaline oxidation of polyphenols and *P. Marfey,* State University of New York, studies the surface structure of yeast cell membranes.

The Department of Physiology

As reported by P. Brandt Rehberg in the preceding chapter, the Carlsberg Foundation had for some time planned to discontinue the Biological Institute of the Carlsberg Foundation by combining it with the Physiology Department of the Carlsberg Laboratory. Accordingly, after the retirement of Holter in 1971, the head of the Biological Institute, Professor *Erik*

Zeuthen, was appointed acting head of the Department of Physiology, in order to prepare the amalgamation of the two Laboratories. However, this plan had to be abandoned, for the reasons related by P. Brandt Rehberg, and in the new situation Professor Zeuthen decided to retain his position as head of the Biological Institute, and offered also to house several coworkers of Professor Holter, and their equipment.

After the transfer of the Department of Physiology to the brewery in 1972, *Diter von Wettstein,* Professor of genetics at the University of Copenhagen, was appointed acting head of the Department of Physiology in order to supervise the appointment of new scientific staff and to inaugurate a series of new research projects.

In 1975, von Wettstein became head of the department. He is now 47 years old. A graduate in botany from the University of Tübingen, Germany, he later became associate professor in genetics at the University of Stockholm, Sweden, and professor of genetics at the University of Copenhagen in 1962.

The Department of Physiology had during the period 1956–72 been primarily occupied with cell physiology. In the new projects, physiological genetics in plants, insects and yeast is emphasized, following up another tradition of the department.

His research interests lie in the fields which connect genetics, cytology, and plant physiology. Accordingly, he has applied experimental techniques from various disciplines and he has demonstrated great ability to master widely varying techniques.

Since v. Wettstein had so recently joined the Carlsberg Laboratory, no evaluation of his previous accomplishments could be given in an earlier, historical, section of this book. It will therefore be attempted here, together with an account of the present activities at the Department of Physiology.

In electron microscopic investigations, v. Wettstein studied the development of chloroplasts in a number of plants, charac-

99 Diter von Wettstein working at the electron microscope.

terizing the ultrastructure of normal types and chlorophyll deficient mutants. These investigations were mainly performed with artificially induced single gene mutants in barley. They led to a scheme describing the development of chloroplasts which can now be found in standard text-books. He has managed to use electron microscopy as a quantitative research tool. He showed that many chlorophyll-deficient mutants were equivalent to biochemical mutants of microorganisms. Together with *T. Nilsson-Tillgren,* v. Wettstein developed a technique permitting synchronous infection of cells in tobacco leaves with tobacco mosaic virus and used this to follow the development of virus particles in the cells.

In recent years he has made contributions to the elucidation of the genetics and ultrastructure of the photosynthetic membranes, considering especially the regulation of the synthesis of their components – proteins, lipids, and pigments. Together with the late Professor *Mogens Westergaard* (Carlsberg Foundation professor at the Department of Chemistry), he studied the synaptinemal complex in relation to genetic crossing-over. Details are given in the obituary on Mogens Westergaard, chapter 4.

Ever since he left Stockholm, v. Wettstein has continued to use the phytotron at the Royal College of Forestry in Stockholm, which was constructed according to his plans. The complete control of the environment afforded by the phytotron enabled him to prove that spruce trees are adapted to their place of growth by a genetically fixed night-length response for budset in the autumn. Arctic trees will terminate their growth when night length exceeds two hours, whereas central European trees will not produce a bud unless the night exceeds eight hours. As one moves from north to south, the genetically determined critical night length changes gradually. v. Wettstein follows up these findings in a cooperative phytotron project which will show whether it is feasible to improve the production of arctic and high-alpine trees by combining the genes for the faster growth of the central European races with the genes necessary for withstanding extreme climates.

v. Wettstein's interests were never confined to basic research in the narrow sense of the word. He has always seen the close connection in plant genetics between pure and applied research and his studies on the induction of mutants in barley with the purpose of improving this grain have to lead to interesting new varieties of barley with greater resistance against bad weather conditions.

One major objective of several projects in the Department of Physiology is to gain insight into some of the mechanisms by which genes regulate the synthesis and function of macromolecular cell structures. With this aim in mind, and using v. Wettstein's extensive single gene mutant collection in barley, the assembly of the photosynthetic membrane in the chloroplast is analyzed. To give an example of the type of approach used, the genetic regulation of chlorophyll synthesis will be discussed in some detail. When the barley grain is allowed to sprout in the dark, it will develop a yellow primary leaf (Fig. 99, right). The chlorophyll biosynthetic pathway has in each chloroplast precursor, the etioplast, produced enough protochlorophyllide to charge the holochrome protein molecules of the etioplast membranes and then the pathway is shut down by repressing the synthesis of ð-aminolevulinic acid. The leaves on the left side of Fig. 99 were also grown in the dark, but are from a mutant in which the regulation of the porphyrin pathway is defect. In the mutant *tigrina-d*^{12}, the porphyrin pathway is not shut down in the dark; instead, ð-aminolevulinic acid and the green pigment protochlorophyllide continue to be synthesized. This is the reason why the leaves are green. As the protochlorophyllide holochrome protein continues to be regulated in the mutant, most of the protochlorophyllide cannot attach to holochrome protein molecules and is therefore not immediately convertible to chlorophyll-a when the leaves are illuminated. When such a regulatory mutant is grown in light-dark cycles, it is tiger-striped (Fig. 99) and lethal. Regulation of chlorophyll synthesis should only be necessary in alternating light and dark cycles and dispensible in continuous light. Fig. 99 shows indeed that the same mutant is fully viable and sets normal seeds when grown in continuous light in the phytotron.

Fig. 100 depicts leaves of the mutant *xantha-l*^{35} which on ð-aminolevulinic acid accumulates considerable amounts of reddish magnesium protoporphyrin and is therefore considered a mutant in the structural gene concerned with a step converting magnesium protophyrin into protochlorophyllide. Combining this structural gene mutant with the regulatory gene mutant discussed above gives seedlings accumulating constitutively magnesium protoporphyrin without supply of ð-aminolevulinic acid. Through these investigations it has be-

100,a Dark grown leaves of wild type barley (right) and the regulatory mutant *tigrina-d*[12] which accumulates large amounts of the green pigment protochlorophyllide, a precursor to chlorophyll.

100,b Progeny segregating for seedlings of the mutant *tigrina-d*[12] (short), seedlings heterozygous for this mutant gene (bent tip), and wild-type seedlings. The plants are grown in alternating light and dark cycles.

come clear that some of the regulatory genes found are concerned with the repression of the porphyrin pathway, whereas others control the feedback mechanisms that are also operating in every biosynthetic pathway. The system is one of the very few cases in which it has been established that genetic regulation of a synthetic pathway in a higher eukaryote proceeds according to principles previously recognized for prokaryotes. *Simon Gough* and *Gamini Kannangara* are at present trying to identify the relevant enzymes for δ-aminolevulinic acid synthesis which are of primary importance in the regulation of the chlorophyll pathway.

The membranes of the etioplasts in the dark grown barley seedlings are concentrated into para-crystalline prolamellar bodies (Fig. 101), which also contain the above mentioned pro-

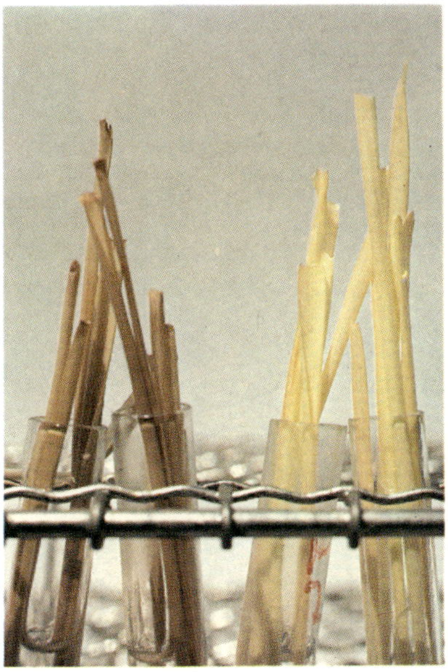

100,c Under continuous light in a phytotron the conditional lethal mutant *tigrina-d*[12] is fully viable.

100,d Leaves of the structural gene mutant *xantha-l*[35]. The leaves on the right have been placed in buffer solution, those on left were provided with δ-aminolevulinic acid. The latter leaves have accumulated the reddish Mg-protoporphyrin.

tochlorophyllide holochrome. Upon illumination and reduction of the protochlorophyllide to chlorophyllide-a, the photosynthetic membranes and grana are assembled as depicted in Figs. 101 and 102.

As the first small grana are built up, new membrane proteins are synthesized differentially and inserted into the pre-existing membranes. Using single gene mutants which in the chloroplast produce different kinds of membrane abnormalities, as illustrated in Fig. 101, something about the function of these proteins has been learnt. Some proteins are responsible for stacking the membrane discs into grana, others prevent stacking. The mutants are further being used in studying which proteins in the membrane belong to the photochemical reaction centers of the chloroplast's two photosystems.

The study of the photosynthetic membrane proteins involves cooperation between *Gunilla Højer-Hansen* and Dr. *Albert Kahn* at Copenhagen University, Professor *Knud W. Henningsen* at the Agricultural University in Copenhagen, Dr. *Otto Machold* at the Centralinstitut für Genetik und Kulturpflanzenforschung in Gatersleben, DDR, Dr. *Niels C. Nielsen* now at the University of California, Davis and Dr. *R. A. Smillie*, CSIRO, Sydney.

Much of the protein contained in the stroma in which the membrane system of the chloroplast is embedded (Fig. 101) consists of the CO_2 fixing enzyme ribulosediphosphate-carboxylase. This large and most abundant plant protein is of special interest as it provides at the molecular level a most beautiful example of the close cooperation of the genetic information in the nucleus and the genetic information contained in chloroplast DNA. According to the studies of Professor *Sam Wildman* at the University of California, Los Angeles, it appears that the small sub-unit of this enzyme is coded for by nuclear DNA and the large subunit by chloroplast DNA. *Steffen Strøbæk, Gregory Gibbons* and *Carsten Poulsen* are engaged in the determination of the primary structure, especially of the small sub-unit of this enzyme from tobacco and barley. This work is carried out in cooperation with Professor *D. Boulter* at the Botany Department of Durham University in England. At the same time work is being done to identify mutants in the small and large sub-units of the enzyme.

Recent electron-microscopic analyses of the developing endosperm in barley by v. Wettstein and Dr. *Lars Munck* have revealed that the storage proteins in this tissue are synthesized and deposited in a way very similar to that worked out originally for the zymogen granules in the pancreas. The protein is synthesized on the ribosomes of the endoplasmatic reticulum, transferred into the lumen of the endoplasmatic reticulum, from there to the Golgi apparatus, and then deposited in the vacuoles of the cells. There the protein precipitates out once a sufficient concentration has been reached. Single gene mu-

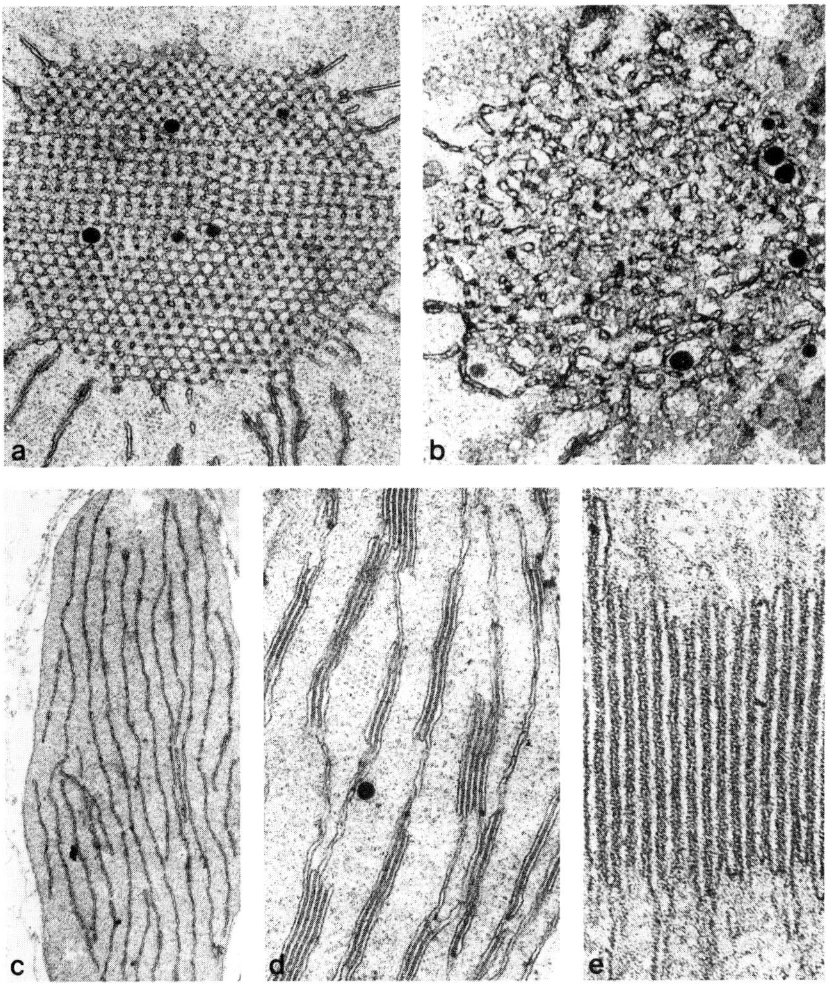

101 Development of an etioplast into a chloroplast in a 7-day-old dark grown barley seedling illuminated with white light. a: Prolamellar body in etioplast. b: Tube transformation: prolamellar body after 1 minute illumination. c: Dispersal of prolamellar body into primary lamellar layers, after 1 minute illumination followed by 15 minutes in darkness. d: Formation of grana: after 24 hours continuous illumination. e: Granum of mature chloroplast. Magnifications, see chapter 14.

tants changing the amino acid composition of the storage proteins are being used to learn more about the composition of these proteins.

John Ingversen and *Anders Brandt* are analyzing the biochemistry of reserve protein synthesis and characterizing the biochemical changes caused by the mutations which change the amino acid composition of the reserve proteins, especially those increasing or decreasing the lysine content. In this project, the Department of Physiology cooperates with the Department of Agricultural Research of the Danish Atomic Energy Commission's Laboratory at Risø, and with Dr. Lars Munck, head of the Carlsberg Research Laboratory, Department of Biotechnology. The latter and the Department of Physiology also share responsibility for the breeding of barley varieties improved with regard to the chemical composition of the grain.

The department works on the biosynthesis of the long-chain lipids which constitute the surface wax of leaves. With the aid of over 1.000 induced *eceriferum* mutants in barley 59 genes have been identified which control the formation and excretion of the wax on the surface of the barley plant. Studies in the phytotron have shown that the wax coating primarily acts as a waterproofing agent. The biochemical genetic investigations are being conducted by *Penny von Wettstein-Knowles* and Dr. *A. G. Netting* from Sydney. *Barbro Strid* uses a series of barley mutants with reduced or absent anthocyanin pigmentation to elucidate the biochemical genetics of this class of compounds in barley.

A major effort of the department is devoted to the analysis of chromosome pairing and genetic recombination during meiosis. In recent work *Denise Zickler* and *L. W. Olson* were able to show that the assembly of the synaptinemal complex and the process of meiosis in yeast proceeds according to the same pattern as has been found in other fungi. *Morten Kielland-Brandt* collaborated during a postdoctoral stay at the

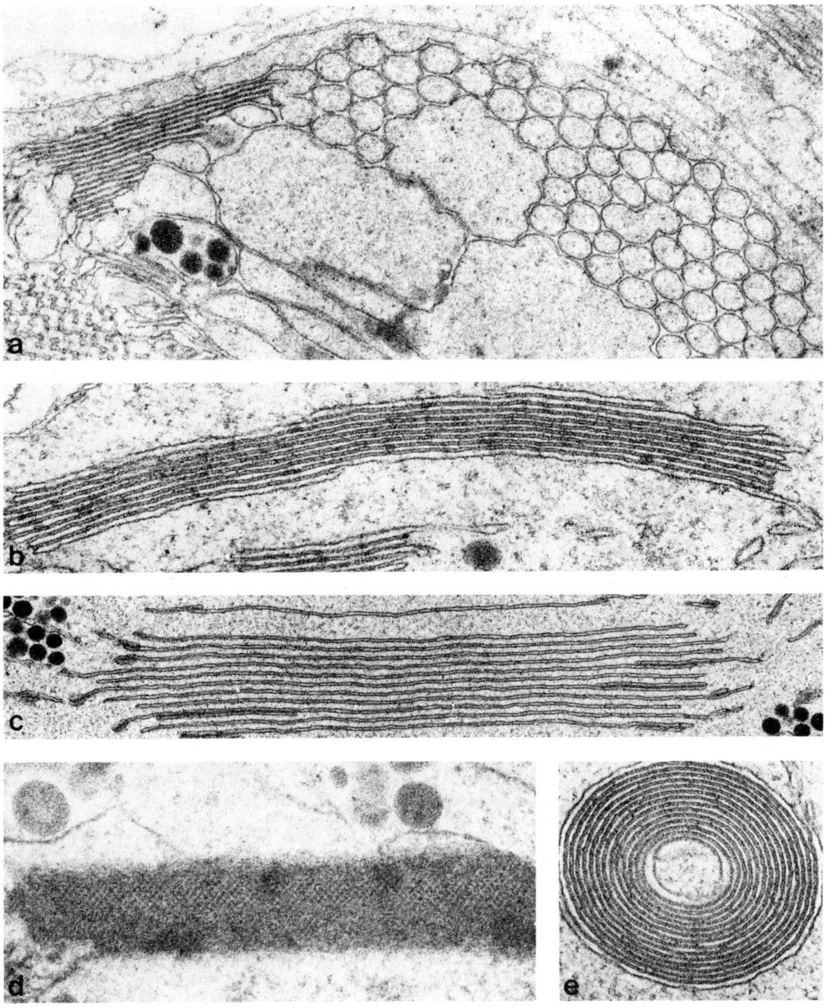

102 Plastid membrane organizations found in five different nuclear gene mutants of barley.

a: *Xantha-a*^{63}: Uncontrolled synthesis of chloroplast specific lipids leads to membranes in honey comb configuration together with grana and prolamellar bodies. b: *Xantha-d*^{31}: Membrane discs (thylakoids) aggregate into giant grana. c: *Xantha-f*^{60}: Mutant blocked in chlorophyll synthesis does not aggregate its discs into grana. d: *Tigrina-o*^{34}: Mutant defective in the regulation of chlorophyll synthesis produces abnormally structured grana. e: *Xantha-b*^{18}: Mutant with spheroidal granum. Magnifications, see chapter 14.

University of Washington in Seattle with Dr. *Walton Fagman* and Professor *Herschel Roman* on the isolation and characterization of chromosomal sized DNA molecules. Kielland-Brandt will be in charge of the section of the department dealing with yeast genetics.

With regard to studies on the biochemistry of meiosis, the department cooperates with Dr. *Yasuo Hotta* and Professor *Herbert Stern* at the University of California, San Diego. In a collaborative project with the Institute of Developmental Biology of the Academy of Sciences of the USSR and especially sponsored by the late director of this institute, Professor *B. Astaurov*, *Søren Rasmussen* has analyzed the female meiosis of the silkworm Bombyx mori. This species is of special interest since the female sex in the silkworm lacks crossing-over. By reconstructing nuclei at various stages of meiosis from electronmicrographs of serial sections, he was able to show that chromosome pairing takes place with the aid of the synaptinemal complex. In the absence of crossing-over, chiasmata which normally hold the repulsing homologous chromosomes together are not formed. Instead, the synaptinemal complex is converted into an elaborate structure that ensures the pairing of the chromosomes up to metaphase I and the coorientation of the bivalents in the metaphase plane. This structure is not shed until anaphase I and remains lying in the equatorial plane of the spindle. The modified synaptinemal complex corresponds to what has been erroneously called elimination chromatin in the classical papers of insect cytology.

The operation of the Carlsberg Laboratory

What is the best framework for successful research activities? Two extreme view points have been expressed: On the one hand there are those who totally deny the importance of the organisation surrounding a laboratory and, on the other hand, those who feel that the organisation is all-important. Concerning basic research, there is a lot to be said for the attitude: »get the best persons and leave them alone«. This has been the attitude of the Carlsberg Foundation towards the Carlsberg Laboratory during most of its existence. The recent transition of the laboratory from being the property of the Carlsberg Foundation to its present status as a part of the United Breweries' Carlsberg Research Center, has raised questions as to the prospects for basic research in the laboratory. This concern was most explicitly voiced in an article in »Science« in March 1972, where doubts were expressed as to the possibility of viable basic research within the new organization. The content of the article – although understandable because of the fluid situation before the final agreement was reached in April 1972 – came as a surprise to those involved in changing the status of the laboratory. In the Carlsberg Foundation we felt strongly that among the many prospects for the laboratory one was certain, namely that it could not continue its existence if it was to be dependent on subventions from the Foundation alone since the expenses would soon exceed the yearly income of the Foundation. The negotiations which led to the incorporation of the laboratory into the brewery research organization are explained in detail by P. Brandt Rehberg in chapter 11.

In discussing this problem there are two important facts to be considered:

Firstly, the laboratory continues to operate under the original statutes formulated by its founder one hundred years ago. Thus its activities are still under the supervision of the Board of Trustees of the laboratory, comprising as previously the three natural science members from the Board of Directors of the Carlsberg Foundation and two adjunct members appointed by the Royal Danish Academy of Sciences and Letters after nomination by the United Breweries. At present *Ebba Lund,* Professor of virology, and *A. W. Nielsen,* the former Executive General Manager of the United Breweries, are serving as adjunct members.

Secondly, it is evident that if outstanding scientists are to be called upon to head the departments of the laboratory, they must feel certain that they will be able to carry on fundamental research of their own choice. If there were any restrictions, it would soon be impossible for the Board of Trustees to find suitable scientists to head the laboratory. The fact that the laboratory to-day is led by two eminent scientists (both members of the Royal Danish Academy of Sciences and Letters) is the best answer to those who might feel uncertain whether the Carlsberg Laboratory will continue its line of basic research. The scientific problems which are worked on to-day and which have been described in this chapter are as fundamental as those studied during earlier periods of the laboratory.

This having been said, it may then be asked: What is the situation with respect to applied research? Again the answer is clear; there is every reason that the Laboratory should also be engaged in research dealing with the application of recently acquired knowledge to the benefit of the breweries. The emphasis put on this aspect of the work has fluctuated over the preceding 100 years. The statutes formulated by the founder, J. C. Jacobsen, leave no uncertainty as to his intentions, namely that the laboratories should serve both purposes. The application of knowledge from the basic research programme to the art of brewing beer is now facilitated by the creation of the Carlsberg Research Center which brings together into a single building the Carlsberg research and development Laboratories and the two original, but extended, departments of the Carlsberg Laboratory.

The Carlsberg Research Center

After the confirmation in April 1972 (se Chapter 11), of the transfer of the Carlsberg Laboratory to the breweries, the brewery management was faced with a number of immediate tasks. First of all, the facilities of the laboratory were to be extended, modernized and supplied with up-to-date equipment. With this objective in view a project was worked out which consisted in a complete renewal of the old laboratory building and the conversion into laboratories of a building hitherto used as a residence for the head of the department of physiology. Since this project was subsequently considered not to comply with the high standards aimed at for a reconstruction, it was discarded, and various possibilities for a new building were considered. In the course of these deliberations the idea arose of collecting together all research facilities in one location. It must not be forgotten that two major breweries had amalgamated and that the research facilities then available were therefore located at various places in Copenhagen.

Based on the strong belief that a fruitful interplay between basic and applied research can advance not only fundamental knowledge, but also the prosperity of the enterprise, the management of the breweries decided to create a large new Research Center adjacent to and linked to the old laboratory buildings. The centre would have as a nucleus the old laboratory, modernized and occupied on both floors by an extended Department of Chemistry, whereas the Department of Physiology and the laboratories for applied research would be located in the new buildings.

A project in conformity with these ideas was approved by the Board of the United Breweries in March 1973. In August 1974 construction work on the site was started and during the summer of 1976 all the laboratories will be built and in operation. At the time of the centenary the new Research Center, notable for its combination of basic and applied research, will be inaugurated. Besides the Departments of Chemistry and Physiology, the Center comprises the Carlsberg Research Laboratory, established by an integration of the existing research laboratories of the Carlsberg and the Tuborg Breweries and various other developmental actives. This laboratory is organized in three departments.

1. The *Department of Brewing Chemistry* devoted to research and development of the chemistry and microbiology of brewing.

This department continues the work carried out in laboratories within the breweries since the late twenties (see Chapter 10). The laboratory is now headed by *B. Ahrenst Larsen,* who for almost 25 years has been engaged in the chemistry of brewing at the Tuborg Breweries. The research carried out in this laboratory ranges from fundamental work on carbohydrates and carbohydrases, especially debranching enzymes, to the chemistry of peptides, polyphenols and hop constituents as well as the fundamental physico-chemical properties of beer.

2. The *Department of Biotechnology* is created as a new unit devoted to research and development within a wide area encompassing the raw materials for brewing and its by-products, together with related topics likely to form the basis of future diversification activities. This laboratory is headed by Dr. *Lars Munck,* who acquired his training in biochemistry, plant genetics and plant breeding at the University of Lund and the Swedish Seed Association at Svalöf. Dr. Munck has been particularly active in the improvement of nutritional values in cereals.

3. The *Department of Brewing Technology* is also a new establishment devoted to developmental work on technology related to the processes of brewing and bottling. Headed by *E. Mernøe,* it is staffed by a small group of engineers who in their previous positions were engaged in mechanical or electronic developments. By now, a number of new inventions are successfully applied in various breweries, e.g. servo control of bottling lines and other automation procedures.

The responsibility for the research and developmental activities of the brewery (R & D) as well as for the construction of the new laboratory building is held by *Eigil Bjerl Nielsen,* who is technical director of the United Breweries (formerly director of research at Tuborg). In passing it may be mentioned that before his appointment at the breweries, Bjerl Nielsen worked as an associate of Professor *John Schellman* in the Department

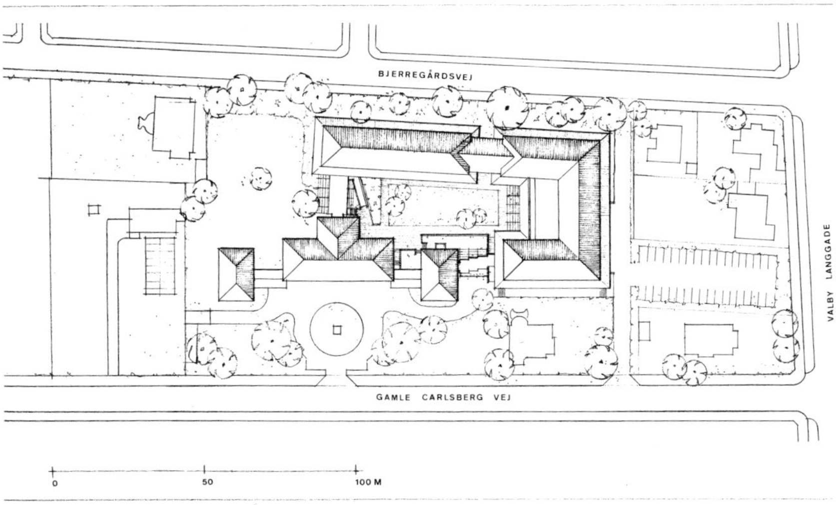

103 Ground plan of the Carlsberg Research Center.

of Chemistry of the University of Oregon. John Schellman had himself been an alumnus of the Carlsberg Laboratory in the last years of Linderstrøm-Lang, and their topic of research (optical properties of amino acids and peptides) originated in the days of Lang at Carlsberg. This may serve as another example of the numerous personal contacts made at the Carlsberg Laboratory. As an aid to the management of research activities an *Advisory Research Council* has been established including as members a number of prominent scientists from outside institutions as well as from within the brewery organization.

The staff of the Carlsberg Research Center comprises about 100 persons, chemists, biochemists, physiologists, geneticists, engineers and technicians, slightly more than half of whom work in the Departments of Chemistry and Physiology. The total laboratory space amounts to about 12.000 m^2 (130.000 sq.ft.) of which the original laboratory buildings comprise 2.600 m^2 (28.000 sq.ft.). Thus a substantial increase in space allotted to research has been provided. Figs. 102 and 103 give an impression of the new buildings.

104 The Carlsberg Research Center during construction, late 1975.

Although organized in five individual departments, all laboratories share auxiliary facilities, such as library, lecture hall, lunch room, workshop, pilot laboratory for large-scale experiments, secretarial assistance and computer facilities. Four apartments are now available for scientists visiting the laboratories for shorter or longer periods.

Besides the laboratories at Carlsberg, facilities are available for field work in plant genetics at the *Carlsberg Experimental Farm,* Allindemaglegård outside Copenhagen, where about 500 acres are at the disposal of the plant breeders for work on barley improvement.

105 The Carlsberg Laboratory 1976.

Epilogue

The reader who has had the patience to reach the last pages of this account of the history of the Carlsberg Laboratory may have noticed the optimism in this final and not so historical chapter. A transition has taken place which has given new impetus to the Laboratory and it is with good hope for the future that the Board of Trustees in 1976 hands over the Carlsberg Laboratory to another 100 years of activity.

In the opinion of the present writer a devoted effort is being made to continue on a large scale and in a modern context the intentions of the founder of the Carlsberg Laboratory. Jacobsen's deep appreciation of the natural sciences and his intuitive and optimistic feeling that industrial endeavour would be furthered in a symbiosis with pure science pervades all those who to-day happen to be responsible for the continued existence of the Carlsberg Brewery and the Carlsberg Laboratory.

XIII

Notes about contributors

The following notes relate mainly to the various contributors' connections with the Carlsberg Laboratory (in case of foreigners with Danish science) and with the purposes of this book. Exhaustive biographical notes have not been attempted.

LISE ALLEN (née Lützen-Nielsen). Typist and proof-reader. Born 1912. Technical assistant at the physiological dept. of the Carlsberg Laboratory 1932–33. Assistant at the Institute of Forensic Medicine, University of Copenhagen 1934–50. Secretary at the chemical dept. of the Carlsberg Laboratory 1950–60. Secretary at the Dept. of Machine Design, Technical University of Denmark since 1961.

RONALD P. BELL. Translator and linguistic adviser. Born 1907. Until 1975 professor of Chemistry, University of Stirling, Scotland. Studied 1928–32 with J. N. Brønsted, Institute of Physical Chemistry, University of Copenhagen. At present visiting professor at the Institute of Physical Chemistry, Technical University of Denmark. Foreign member of the Royal Danish Academy of Sciences and Letters (1962); Dr. techn. h. c., Technical University of Denmark.

CHR. CRONE. Author of chapter 12. Born 1926. Professor of Medical Physiology, University of Copenhagen (1967). Member of the board of the Carlsberg Foundation since 1971. Present chairman of the board of trustees of the Carlsberg Laboratory.

VILH. EGE. 1887–1962. Author of Johannes Schmidt's biography, chapter 5 (obituary in Compt. Rend. Trav. Lab. Carlsberg, vol. 20, (1933–1934) I-IX (1935), with slight changes). Assistant at the physiological dept. of the Carlsberg Laboratory 1921–33. Scientific associate of the Marine Biological Laboratory, Copenhagen 1933–57.

BENT FOLTMANN. Author of M. Ottesen's biography, chapter 4. Born 1925. Professor of Biochemical Genetics, University of Copenhagen. Assistant at the chemical dept. of the Carlsberg Laboratory 1959–63.

DAVID GLICK. Author of one of the contributions to chapter 8. Born 1908. Director of the Center for Histochemical Research, Stanford Research Institute, California. Visiting scientist at the Carlsberg Laboratory 1933–34, 1936–37, 1949, 1958, and numerous shorter visits.

H. HOLTER. Editor, author of chapters 1, 3 and 7, a contribution to chapter 8, and K. Linderstrøm-Lang's biography in chapter 4 (obtituary in Compt. Rend. Trav. Lab. Carlsberg, vol. 32, (1960–1962) I-XXXIII (1962), abridged). – For personal data see chapter 5.

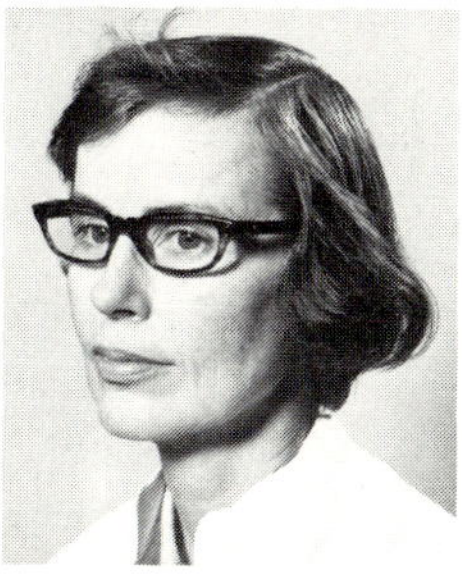

KAREN HOLTER (née Teisen). Typist and editorial assistant. Born 1914. Technical assistant in chemical (1937–43), cytochemical (1943–56), and physiological depts. of the Carlsberg Laboratory. Now retired.

ROLLIN D. HOTCHKISS. Author of one of the contributions to chapter 8. Born 1911. Professor of Cellular Physiology, The Rockefeller University, New York. Visiting scientist at the Carlsberg Laboratory 1937–38, and several shorter visits.

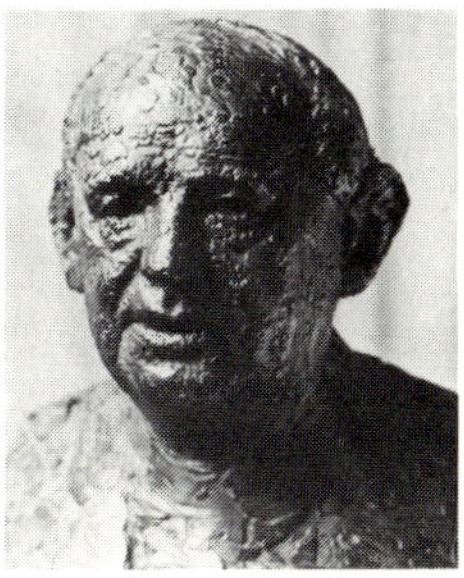

A. TOVBORG JENSEN. Author of chapter 6.
Born 1911. Niels Bjerrum's successor as Professor of Chemistry, Veterinary and Agricultural University, Copenhagen (since 1949). Member of the board of the Carlsberg Foundation 1956–72. Chairman of the board of trustees of the Carlsberg Laboratory 1960–72.

W. JOHANNSEN. 1857–1927. Author of J. Kjeldahl's biography in chapter 4 (obituary in Meddelelser fra Carlsberg Laboratoriet, vol. 5. (1900–1903), I–XI (1903), from the Danish version with slight changes). Professor of Plant Physiology, University of Copenhagen (1905–27). Assistant at the chemical dept. of the Carlsberg Laboratory 1881–87.

H. M. KALCKAR. Translator and author of a contribution to chapter 8. Born 1908 in Copenhagen, professor of Biological Chemistry, Harvard University Medical School, Boston. Has been director of Institute of Cytophysiology, University of Copenhagen (1949–56). Emigrated to USA in 1956. Visiting professor at the Carlsberg Laboratory on several occasions.

JØRGEN KIELER. Author of a contribution to chapter 8. Born 1919. Director of the Fibiger Laboratory (Cancer Research) in Copenhagen since 1960. Visiting scientist at the Carlsberg Laboratory 1955–56.

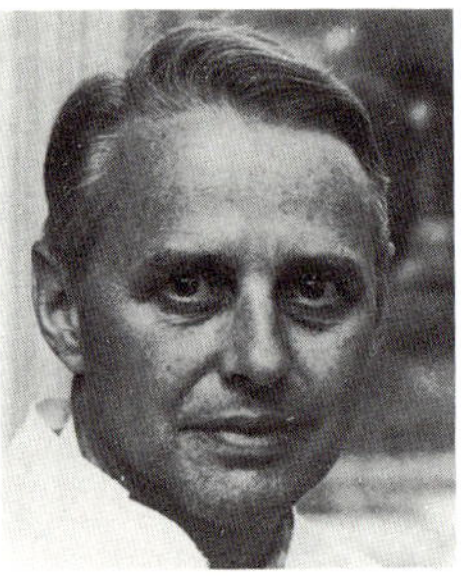

A. KLÖCKER. Author of E. Chr. Hansen's biography in chapter 5 (obituary in Meddelelser fra Carlsberg Laboratoriet, vol. 9 (1910–1913), I–XXXVI (1913), from the Danish version, with abbreviations). – For personal data see biography in chapter 5.

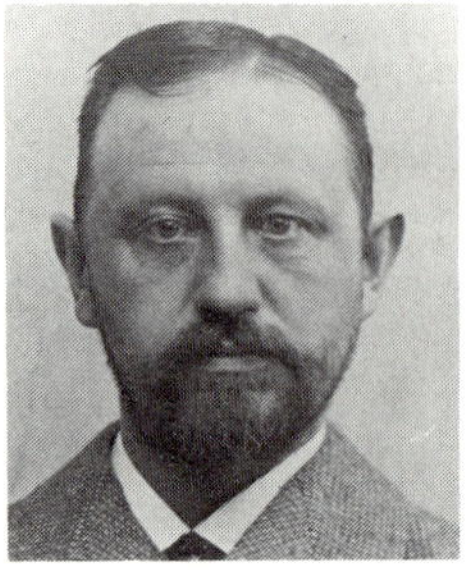

R. KOEFOED. 1862–1924. Author of H. Schjerning's biography in chapter 4 (obituary in Meddelelser fra Carlsberg Laboratoriet, Vol. 11 (1914–1917) 351–358 (1917), from the Danish version, abridged). Assistant at the chemical dept. of the Carlsberg Laboratory 1886–90. Active in management of Gamle Carlsberg Brewery 1892–1901, director 1901–06. Member of the board of trustees of the Carlsberg Laboratory 1906–24.

POUL LARSEN. Author of C. Olsen's (chapter 4) and N. Nielsen's (chapter 5) biographies. C. Olsen's biography is based on an obituary in Overs. Dan. Vid. Selsk. 1974–75, 109–122, (1975) much abridged and with a few additions taken from E. Bille Hansen's obituary in Bot. Tidsskr., vol. 79, 98–100 (1975). Born 1909. Professor of Plant Physiology, Aarhus University. Deceased July 8, 1976.

K. LINDERSTRØM-LANG. Author of S. P. L. Sørensen's biography in chapter 4 (obituary in Compt. Rend. Trav. Lab. Carlsberg, Sér. Chim., vol. 23, I-XXI (1941) abridged), and of another contribution to the same chapter (from: European Brewery Convention, Proceedings of the Congress, Copenhagen 1957. Elsevier Publishing Company (1957), pp. 4–10). – For personal data see chapter 4.

K. MAX MØLLER. Editor of illustrations and historiographer. Born 1922. Scientific volunteer at chemical dept. (1944–45), assistant at chemical (1949–54), cytochemical (1954–56), and physiological (1956–72) depts. of the Carlsberg Laboratory. Now head of division at the Biological Institute of the Carlsberg Foundation.

DETLEV MÜLLER. Author of R. Pedersen's biography in chapter 5. Born 1899. Professor of Plant Physiology, University of Copenhagen 1949–69. Now retired.

E. STEEMANN NIELSEN. Author of chapter 9.

Born 1907. Professor of Limnology, University of Copenhagen since 1969. Has participated in a number of oceanographic expeditions, among them Johannes Schmidt's »Dana« expedition 1928–30 and the first part of the »Galathea« expedition 1950–52.

P. BRANDT REHBERG. Author of chapter 11.

Born 1895. Professor of Zoophysiology, University of Copenhagen (1945–65). Member of the board of the Carlsberg Foundation 1960–71, chairman 1969–71, member of the board of trustees of the Carlsberg Laboratory 1961–70. Now retired.

CHARLOTTE SCHELLMAN (née Green). Linguistic adviser.

Born 1922. Adjunct Associate Professor of Chemistry, Institute of Molecular Biology, University of Oregon, Eugene, Oregon. Visiting scientist at the Carlsberg Laboratory 1953–55 and numerous shorter visits.

JOHN A. SCHELLMAN. Linguistic adviser.
Born 1924. Professor of Chemistry, Institute of Molecular Biology, University of Oregon, Eugene, Oregon. Visiting scientist at the Carlsberg Laboratory 1953–55 and numerous shorter visits.

H. SCHIÖNNING. 1868–1942. Author of A. Klöcker's biography in chapter 5 (obituary in Meddelser fra Carlsberg Laboratoriet, vol. 15, (1923–1925), I–VI (1925), from the Danish version, with slight changes). Assistant at the physiological dept. of the Carlsberg Laboratory 1894–1909, afterwards official in the beer taxation service.

B. TROLLE. Author of chapters 2 and 10.
Born 1910. Since 1937 active in the laboratories of the Carlsberg Breweries; head of the Research Laboratory of the Carlsberg Breweries 1946–73. Since 1973 head of the Central Laboratory of United Breweries Ltd. Member of the European Brewery Convention's Barley Committee (since 1949), and chairman for its Hop Liaison Committee (since 1959).

L. VANDENDRIESSCHE. Author of a contribution to chapter 8.
Born 1919. Professor of Biochemistry at the University of Ghent. Visiting scientist at the Carlsberg Laboratory 1947–50 and numerous shorter visits. At present rector of the University of Antwerp.

MOGENS WESTERGAARD. 1912-1975. Author of Ø. Winge's biography in chapter 5 (obituary in Compt. Rend. Trav. Lab. Carlsberg, vol. 34, (1963–64), I-XXIV (1965), abridged and slightly corrected). – For personal data see chapter 4.

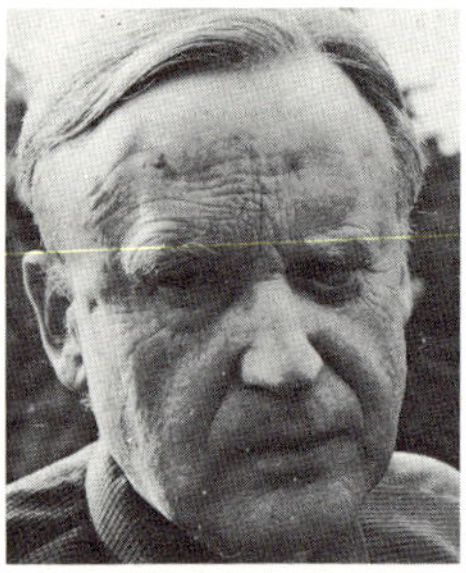

D. von WETTSTEIN. Author of M. Westergaard's biography in chapter 4. Born 1929. Present head of the dept. of physiology at the Carlsberg Laboratory. – For additional data see chapter 12.

ERIK ZEUTHEN. Author of H. Holter's biography in chapter 5. Born 1914. Assistant at the chemical and cytochemical depts. of the Carlsberg Laboratory 1939–46. Head of the Biological Institute of the Carlsberg Foundation since 1957; professor since 1961.

XIV

Selected references, acknowledgements and comments on illustrations

SELECTED REFERENCES

Carlsbergfondets beretninger
In: Overs. Dan. Vid. Selsk. (Annual in Danish)
From the volume covering the year 1876, Kjøbenhavn (1876-78) (with a French summary) until the volume covering June 1974 to May 1975, København (1975) (with an English summary).

Carlsbergfondets Direktion (udgivet af)
Carlsbergfondet 1876-1926 Et Jubilæumsskrift
Vols I/II
Bianco Lunos Bogtrykkeri A/S,
København (1930)

Meddelelser fra Carlsberg Laboratorium (and Compt.Rend.Trav.Lab. Carlsberg) from vol. 1 (1878) to vol. 40 (1976).

Christensen, C.
Den Danske Botaniks Historie med tilhørende Bibliografi
vols: I^1, I^2 & II.
København: H. Hagerups Forlag (1924–1926)

Engelstoft, P. and S. Dahl (Red.)
Dansk Biografisk Leksikon
Vols. 1–27 (1933–1944)
J. H. Schultz Forlag, København.

Fraenkel, A.
Gammel Carlsberg
H. Hagerups Boghandel, København (1897).

Holter, H.
Laboratories of the Carlsberg Foundation
Research 3, 275–279 (1950).

Kraks Blå Bog
(Annual Danish »Who's Who«)
Krak, København 1910 ff.

Nyrop, C.
J. C. Jacobsen 1811–1911. Et mindeskrift.
Kjøbenhavn (1911).

Pedersen, J.
The Carlsberg Foundation
Copenhagen (1956).

Schmidt, I.
Blade af Carlsbergs Historie.
Bianco Lunos Bogtrykkeri A/S, København. (1956).

Veibel, Stig
»Kemien i Danmark«.
Vol. I (1939) Vol. II (1943) Vol. III (Bodil Jerslev, Editor; 1968)
Nyt Nordisk Forlag, Arnold Busck, København.

Winge Ø. and Å. Vedel Tåning
Naturforskeren Johannes Schmidt. Hans Liv og Ekspeditioner Skildret af Venner og Medarbejdere.
København: Gyldendalske Boghandel, Nordisk Forlag (1947).

ARCHIVES AND MUSEUMS

The Carlsberg Foundation
35, H. C. Andersens Boulevard 35
DK–1553 Copenhagen V.

The Carlsberg Foundation Picture Archives
(Abbreviated C.F.P.A.)
c/o Carlsbergfondets Biologiske Institut,
16, Tagensvej 16
DK–2200 Copenhagen N.

The Carlsberg Laboratory
10, Gl. Carlsbergvej 10
DK–2500 Valby

The Carlsberg Museum
1, Valby Langgade 1
DK–2500 Valby

The Danish Institute for Fishery and Marine Research
Charlottenlund Slot
DK–2920 Charlottenlund

The Danish National Archives
9, Rigsdagsgaarden 9
DK–1218 Copenhagen K.

The Museum of National History at Frederiksborg Palace
DK–3400 Hillerød.

The Royal Library
8, Christians Brygge 8
DK–1219 København K.

ACKNOWLEDGEMENTS AND COMMENTS ON ILLUSTRATIONS

In the notes of this chapter the editors have tried to give information about the illustrations of this book, so as to enable an interested reader to find the sources used. On the other hand, we had to avoid making the chapter too cumbersome, and we are therefore well aware that some relevant informations and acknowledgements may be missing. We can only state that all such omissions are quite unintentional, also in cases where they might be relevant to questions of copyright.

In the description of illustrations all dimensions are given in cm or mm, the height before the width. The dimensions are those of the actual picture surface, without the frame. The terms »left« and »right« refer to the spectator's left and right, except otherwise indicated. The expression »detail« denotes any area of a picture smaller than its total area. The nationality of artists is only mentioned if they are not Danish.

The abbreviation »C.F.P.A.« denotes that the illustration mentioned is to be found in the Carlsberg Foundation Picture Archives.

Frontispiece Jacob Christian Jacobsen (1811–1887), oil on canvas, 166 by 111,5 cm, by August Jerndorff. Inscribed below on the left: Carlsberg Dec 1886 A.J.
The Museum of National History at Frederiksborg Palace. Inventory No. A 2644. Copies made by the artist himself in 1886 and 1887 are known (Carlsberg Foundation office and Carlsberg Breweries administration office).
Photography: Lennart H. F. Larsen, Copenhagen; 12 by 9 cm colour diapositive.
By courtesy of The Museum of National History at Frederiksborg Palace.

1 Bird's-eye view of Gammel Carlsberg brewery with the Annex brewery (the former first Ny Carlsberg), in the foreground in 1883.
Drawing and wash by the architect C. F. Thomsen. Taken from a mounted photograph 33.9 by 50.8 cm at the Carlsberg Museum. Inventory No 3720. These photographs were distributed, under the title »1847–1883«, to the

friends of Carlsberg in commemoration of the events that happened at the Carlsberg Laboratory in 1883 (introduction of pure yeast and the Kjeldahl method). C.F.P.A. No. 1976–342.

2 Front view and plan of the first Carlsberg Laboratory.
Drawing by architect C. F. Thomsen. Taken from Meddelelser fra Carlsberg Laboratoriet *5* (1900–1903), Extra Hefte (1901), p. 49 Pl. I., text from pp.3–5. The dotted sector on the plan indicates the field of vision shown in the painting of J. Kjeldahl (figure 8). C.F.P.A. No. 1976–343.

3 The Carlsberg Laboratory of 1897. Drawings by architect C. F. Thomsen. Taken from Meddelelser fra Carlsberg Laboratoriet *5* (1900–1903), Extra Hefte (1901) p.49., Pl. III, V & Vl; text from pp. 10–16.

4 Changing population density, I. Chemical department 1897. Taken from a photograph at the Carlsberg Museum, 17 by 23 cm. Photographer: unknown. C.F.P.A. No. 1964–283.

5 Changing population density, II. (1909). Taken from a photograph at the C.F.P.A., No 1964–376, 18 by 24 cm. Photographer: C. Ferslew & Co., Copenhagen. The photograph was taken on the occasion of the H. C. Ørsted medal beeing conferred on S. P. L. Sørensen.

6 Changing population density, III. (1958). Taken from a photograph published in the German weekly »QUICK«, 11, No 8 February 1958. Photographer: Hilmar Pabel. Copy from original in the C.F.P.A., No. 1964–313, 21 by 29 cm. Persons viewed from left to right: 1. Charles C. Bigelow, 2. Kirsten Mikkelsen, 3. A. T. Ganesan, 4. Paul Ottolenghi, 5. Heinz Holter, 6. Philip Wilcox, 7. M. Funatsu, 8. Sigurd Olaf Nielsen, 9. Earl Frieden, 10. James Cowey, 11. Joseph L. Rabinowitz, 12. Jytte Hjo Taplov, 13. Ellis S. Benson, 14. Stella Reichstein Sørensen, 15. K. Linderstrøm-Lang.

7 The building Gl. Carlsbergvej No. 8. Photo: Johannes Hansen, 17. March 1976. Detail of 24 by 36 mm negative. C.F.P.A. No. 1976–57.

8 Johan Kjeldahl in the old Carlsberg Laboratory (see also figure 2). Oil on canvas, 157 by 220 cm by Otto Haslund. Inscribed below on the right: Otto H. 1897. Two existing photographs of Johan Kjeldahl taken in Ludvig Grundtvig's studio have been used in the portrait painting. The painting is situated on the northern wall of the assembly room of the Carlsberg Laboratory. It was a donation from the Gl. Carlsberg brewery: »as a visual memory of the first modest home of the laboratory...«. Photographed by Hans Petersen, The Royal Museum of Fine Arts, Copenhagen; 13 by 18 cm colour diapositive. C.F.P.A. No. D75–1.

A sketch, oil on canvas, 54 by 74 cm by Otto Haslund, inscribed below on the right: Otto H., is in The Museum of National History at Frederiksborg Palace. (Inventory No. A4455). A copy, oil on canvas, 79.4 by 110.4 cm by Rasmus Christiansen, inscribed below on the right: R x CHR, 1927, Kopi efter Otto H. 1897, is in the Carlsberg Museum, (Inventory No. 3368).

9 The original Kjeldahl apparatus for the determination of nitrogen. Taken from a woodcut 12 by 7 cm, by J. F. Rosenstand, as published in Meddelelser fra Carlsberg Laboratoriet *2* (1883–1888) p. 331, and in the French résumé p. 198. C.F.P.A. No. 1975–530.

10 »Kjeldahl in his remarkable and genuine modesty«. Detail of a photograph at the Carlsberg Museum (13 by 17.6 cm). Photographer: unknown. Another copy with a wider view exists in an album belonging to Mrs Eva Weis Bentzon. C.F.P.A. No. 1964–275.

11 S. P. L. Sørensen, 58 years old. Oil on canvas 94 by 74 cm by Ludvig Find. Inscribed below on the right: Find 1926. The painting is situated on the western wall of the assembly room of the Carlsberg Laboratory, above the door. Photographed by Hans Petersen, The Royal Museum of Fine Arts, Copenhagen; 12 by 9 cm colour diapositive. C.F.P.A. No. D75–2.
A copy, oil on canvas 94.4 by 73.4 cm by Knud Edsberg (1969) is situated in the Carlsberg Museum: Inscribed below on the right: Find 1926, Edsberg.

12 Sørensen at the time of the amino acid syntheses (about 1909). Detail of a photograph (17.7 by 12.8 cm) on printing-out paper, belonging to a picture collection donated by Niels Berg to the H. C. Ørsted Institute in Copenhagen. Photographer: unknown. C.F.P.A. No 1966–166. A nearly identical picture was published in the weekly: SORØ AMTSTIDENDES SØNDAGSLÆSNING No. 47, Sunday 21. November 1909.
C.F.P.A. No. 1964–40.

13 Effect of the addition of acid on the activity of invertase. Redrawn from: Compt.Rend.Trav.Lab. Carlsberg *8* (1909) figures in the tables No. 16, 17, 18 and 19, on pages 126–129, and figure 2, page 132.

14 Chart of the Sørensen buffer system. Redrawn from: Bertho, A. and W. Grassmann: Laboratory Methods of Biochemistry, Macmillan & Co., Ltd. London (1938), chart in the back of the book. This chart is a simplified version of S. P. L. Sørensen's original, published in: Compt.Rend.Trav.Lab. Carlsberg. *8* (1909), folded diagram inserted after page 174 (50 by 70 cm).

15 Egg Albumin crystals. Taken from a photograph belonging to an album produced by H. Holter and O. C. Richardt and presented to S. P. L. Sørensen on occasion of his seventieth birthday l. January 1938. Detail of a 13 by 18 cm picture, C.F.P.A. No. 1964–175.

16 Sørensen crystallizing the protein Egg Albumin by ammonium sulphate precipitation. (1936). The photograph dates back to the weekly magazine »UGE-JOURNALEN« (»VECKO–JOURNALEN«) Sunday 19. April 1936, No. 16.p.10. Photographer: unknown.
Taken from a copy 22.5 by 16.5 cm in the C.F.P.A. No. 1963–30. Verso: S. P. L. Sørensen's signature and the date 23/4 1936 in his handwriting. A written statement by K. U. Linderstrøm-Lang on the verso of another copy (No. 1963–31): »This is according to my opinion the best picture of Sørensen«. Linderstrøm-Lang used it for an oil on canvas painting belonging to Gerda Linderstrøm-Lang.

17 Sørensen titrating, using the original set of buffer stock solutions and indicators (1937). Taken from a photograph at the Carlsberg Museum, inventory No. 347, C.F.P.A., No. 1964–43, 24 by 17.3 cm. The photograph dates back to the weekly magazine »UGE-JOURNALEN« (»VECKO-JOURNALEN«) Sunday 19. April, No. 16.p.10. Photographer: unknown.
Notice: Below the opal-glass covered table, the top part of the electrolytic hydrogen generator is visible. The hydrogen was piped via a long copper tube into the adjacent room with a large air thermostat which contained the hydrogen electrodes during measurements of pH (the »doll's house«).

18 The only available picture of H. Schjerning, taken in his later years. An elliptically shaped (10.1 by 14.2 cm) mounted photograph borrowed from Chr. Wanning, by courtesy of Mogens Westergaard. Photographer: unknown. C.F.P.A. No. 1972–152.

19 K. Linderstrøm-Lang.
Detail from a 36 by 24 mm negative film (No 31). Photographed by C. F. Jacobsen in the garden of the Carlsberg Laboratory some time in autumn 1939. C.F.P.A. No. 1975–379.

20 Linderstrøm-Lang, 15 years of age, with his sister Karen in the garden of the family home at Kong Georgsvej 45, Copenhagen. Detail from a photograph 7.8 by 10.9 cm belonging to Gerda Linderstrøm-Lang. Photographer: unknown. C.F.P.A. No. 1964–184.

21 The original Carlsberg microtitration outfit.
Taken from a photograph 14 by 10 cm. Photographer: unknown. Possibly distributed by glass blower F. C. Jacob, Copenhagen. Inscribed above on the right 103 and below on the right 30498. C.F.P.A. No. 1964–314.

22 Micro-dilatometric technique.
Taken from Kaj Ulrik Linderstrøm-Lang: Lane Medical Lectures. Proteins and Enzymes. Stanford University Press, Stanford California (1952) p.11 Figure 6. C.F.P.A. No. 1975–531.

23 Linderstrøm-Lang's model of the hog insulin molecule, built from »Catalin Molecular Covalent Models« (1 cm = 1 Å). Detail from a photograph by M. Ottesen, January 1955, 24 by 36 mm film.
Taken from an enlargement 12.7 by 23.4 cm. C.F.P.A. No. 1964–863.

24 Linderstrøm-Lang performing a deuterium exchange experiment. Photographer: F. Aggersbo (©Foto-service I/S, Copenhagen). Exposure 8. December 1951 on 9 by 12 cm plate. (No. S4344). C.F.P.A. No. 1964–381.

25 Posthumous portrait of K. Linderstrøm-Lang. Oil on canvas 105 by 91 cm by Knud Agger. Based mainly on two black and white photographs (C.F.P.A. No. 1964–197 and 1964–183) and on the features of Linderstrøm-Lang's son Carl Ulrich Linderstrøm-Lang who sat for the painting, and the facial colours of his daughter Birte Max Møller; finished November 1960. The painting is situated on the northern wall of the staircase of the Carlsberg Laboratory. Photographed by Hans Petersen, The Royal Museum of Fine Arts, Copenhagen, 12 by 9 cm colour diapositive.
C.F.P.A. No. D75–3.

26 One of Linderstrøm-Lang's many talents. He is entertaining a party given by Charlotte Green (later Schellman) at that time a visiting scientist at the Carlsberg Laboratory (springtime 1953). Photographed by Henrik Ege. Detail of a 24 by 36 mm negative. C.F.P.A. No. 1964–378.

27 Linderstrøm-Lang and his wife Gerda in gay mood facing their old friend David Glick. Photographed by David Glick (U.S.A.) in front of the staircase leading from the diningroom to Linderstrøm-Lang's garden, July 1949. Detail of a 10.4 by 6.1 cm positive. C.F.P.A. No. 1975–308.

28 Carsten Olsen in 1939. Photographed by C. F. Jacobsen in the Carlsberg Laboratory garden. Detail of a 36 by 24 mm negative (No. 32). C.F.P.A. No. 1975–382.

29 Olsen performing an experiment on pH variation in plant water cultures (1917). Detail of a photograph made by Carsten Olsen himself on 1. June 1917, 9 by 12 cm glas negative. Here taken from a positive 8.3 by 11.3 cm on printing-out paper; by courtesy of Elsa Olsen. C.F.P.A. No. 1976–105.

30 Water culture experiment with wavy hair-grass (Deschampsia flexuosa). Photographer: unknown. Detail of a 9 by 12 cm colour diapositive. (By courtesy of Elsa Olsen). Published in Compt.Rend.Trav.Lab. Carlsberg *31* (1960) Plate I, Page 59.

31 Olsen in his laboratory (1936). The photograph dates back to the weekly Magazine, »UGE-JOURNALEN« (»VECKO-JOURNALEN«) Sunday 19. April 1936, No. 16. p.11. Photographer: unknown. Taken from an original copy 16.5 by 22.4 cm. C.F.P.A. No. 1963-131.

32 Linderstrøm-Lang's Cervantesque caricature of C. Olsen and G. Hevesy.
Drawn for the Christmas party 1937. Crayon on cheap paper (originally a roll 94 cm wide) now cut to 66 by 47 cm. Contour and Olsen's face: black. The laurel wreath: green. Hevesy's moustache and hair: light blue. This drawing is deposited together with all other Linderstrøm-Lang drawings in the Royal Library, Copenhagen, Department of Maps and Pictures, as No. 55. C.F.P.A. No. 1964–327.

33 Mogens Westergaard (1955). Detail from a photograph 12 by 9 cm negative film by J. Juncker-Jensen, Copenhagen. (Negative No. 13418–C; 22. June 1955). Copied by the Royal Library, Copenhagen, Department of Maps and Pictures (No. 102806). C.F.P.A. No. 1975–330.

34 Westergaard and his co-worker O. Frydenberg in discussion (1959). Photographed at the Institute of Genetics, University of Copenhagen in January 1959, by Walter Månsson (© Nordisk Pressefoto A/S, Copenhagen). Used on several occasions in the Newspaper BERLINGSKE TIDENDE. Taken from a positive 24.9 by 20.5 cm. C.F.P.A. No. 1975–344.

35 Orchids in a meadow, photographed by M. Westergaard. Taken at Høvblege on the island of Møn in 1972. From a colour diapositive 24 by 36 mm. By courtesy of Ebba Westergaard.

36 Martin Ottesen (1976). Detail from a photograph taken at the Carlsberg Laboratory on 21. January 1976 by Rigmor Mydtskov, Copenhagen, on 36 by 24 mm film. C.F.P.A. No. 1976–160.

37 Ottesen at his home-built fraction collector for the analysis of peptides (1951). Photographed in his laboratory by Aggersbo (©Foto-Service I/S, Copenhagen) 8. December 1951 on 12 by 9 cm negative plate No. S4346. Detail from a positive 23.3 by 17.1 cm. C.F.P.A. No. 1975–329.

38 The first throughly studied example of limited proteolysis. Taken from Compt.Rend.Trav.Lab. Carlsberg Sér.Chim. *30* (1958) p. 262.

39 Subtilisin. Model showing the three-dimensional structure of a molecule of the crystalline enzyme Subtilisin Type NOVO. From a 12 by 9 cm, colour diapositive made by John Carrebye, Copenhagen September 1969. By courtesy of NOVO Industri A/S.

40 Rasmus Pedersen (1904). Detail from a mounted and framed photograph on printing-out paper 14.8 by 10.0 cm by Eugenius Warming, August 1904, according to the handwriting of W. Johannsen on the verso of the frame. But according to Carl Christensen: »Dansk Botanisk Litteratur 1880–1911« Copenhagen (1940), it is by Knud J. Andersen, Rønde, 1905. It is taken outside the Botanical Laboratory at the University of Copenhagen.By courtesy of Professor E. G. Jørgensen, Institute of Plant Physiology, University of Copenhagen. C.F.P.A. No. 1976–136.

41 E. Chr. Hansen in the old Carlsberg Laboratory. Oil on canvas, 157 by 220 cm by Otto Haslund. Inscribed below on the left: Otto H. 1897.
On the left sitting: E. Chr. Hansen; on the right standing, the laboratory steward, Peter Andersen. One existing photograph of E. Chr. Hansen taken in the old Carlsberg Laboratory by Ludvig Grundtvig has been used in the portrait painting. The painting is situated on the southern wall of the assembly room of the Carlsberg laboratory. It was given to the laboratory by the Carlsberg Brewery: »as a visual memory of the first modest home of the laboratory . . .«.
Photographed by Hans Petersen, The Royal Museum of Fine Arts, Copenhagen; 13 by 18 cm, colour-diapositive. C.F.P.A. No. D 75–6.
A sketch, oil on canvas, 54 by 74 cm by Otto Haslund, inscribed below on the right: Otto H., is in The Museum of National History at Frederiksborg Palace (Inventory No. 3434).

42 The young E. Chr. Hansen (c. 1875). Taken from Carl Christensen: Den Danske Botaniks Historie: København (1924–26) vol. I. p. 719, fig. 184, 7.5 by 5.1. cm. Photographer: unknown. C.F.P.A. No. 1966–162.

43 Apparatus for propagation of pure yeast from single cells, as used in the Ny Carlsberg Brewery in 1888 (compare figure 93). Taken from a photograph of a woodcut executed by Th. Lange in 1889. The photograph 15.9 by 16.1 cm belongs to the Carlsberg Museum. C.F.P.A. No. 1975–398.

44 Yeast cells drawn by E. Chr. Hansen. Taken from Meddelelser fra Carlsberg Laboratoriet *2* (1883–1888) Tab. I after p. 210, text p. 209 or in French résume p. 47 (1888).
A Copper plate 20.7 by 132 cm, engraved by E. A. Løvendal. In this copy which belonged to Emil Chr. Hansen personally, he has added in pencil the notes S. cerevis. I., S. Past. I and S. Past II. C.F.P.A. No. 1976–351.

45 Hansen at his microscope in the new Carlsberg Laboratory (c. 1900). Foreground: The »Hansen box« for the sterile handling of yeast. Background: Hansens office with desk. Detail from an original mounted photograph 16.5 by 22.5 cm, on printing-out paper by Fred. Riise. Inscribed below on the cardboard Fred. Riise fot. København K. Vimmelskaftet 42. By courtesy of the Carlsberg Museum. C.F.P.A. No. 1975–397.

46 A. Klöcker (1862–1923). Detail of a framed photograph 12 by 9 cm on printing-out paper by the amateur photographer and botanist Henning E. Petersen. The other person in the background on the right is unidentified.
Inscription on *verso*: Denne kopi er tagen af Klöcker selv. Pladen overlod jeg til Klöcker. HEP. H.E.P. fot. Sept. 1907. By courtesy of S. L. Tuxen, the Zoological Museum, University of Copenhagen. C.F.P.A. No. 1966–164.

47 Johs. Schmidt (1916). Original clay model for marble medallion at the Carlsberg Laboratory. This elliptic portrait medallion by the Swedish artist Ida C. Thoresen is only known from this photograph (negative on glass plate 18 by 13 cm; Photographer: unknown) and from three actual copies: A plaster copy at the Carlsberg Museum (49.0 by 39.8 cm; Inventory No. 3361) inscribed Ida C. Thoresen, 1961; a copy carved in marble at The Museum of National History at Frederiksborg Palace (48.0 by 39.4 cm; inventory No. A 4839), inscribed Ida C. Thoresen 1934, and the one also of marble in a marble frame sunk into the wall at the top of the staircase in the Carlsberg Laboratory (48.0 by 39.5 cm; inscribed Ida C. Thoresen, 1935). C.F.P.A. No. 1976–258.

48 Harvest of hops (1912) in the greenhouse established by Johs. Schmidt for hop breeding at the Carlsberg Laboratory (1911). Detail from a 18 by 13 cm negative plate by Ø. Winge, 12. Sept. 1912. The names of the two ladies and the gentleman are unknown. C.F.P.A. No. 1976–235.

49 Schmidt on board the »Dana« II (1922) with a ring-net of his own construction. Taken from a photograph on printing-out paper 8 by 13 cm by an unknown photographer, January 1922. The picture belongs to an album in the Danish National Archives (Carlsbergfondets Dana Ekspedition Arkiv nr. 10.291 pk. nr. 6). C.F.P.A. No. 1975–283.

50 Johs. Schmidt and his collaborators, the botanist Ove Paulsen (left) and the hydrographer J. N. Nielsen (right). »Dana« II at Gibraltar 1928. Detail of a photographic negative plate 9 by 12 cm, taken by an unknown photographer on board »Dana« II in the end of July 1928 in commemoration of a similar picture taken 25 years earlier under Iceland on board »Thor«. Found in the »Dana« collections at Charlottenlund Slot.
C.F.P.A. No. 1976–44.

51 Reception at the Carlsberg Laboratory on the occasion of a meeting of the International Council for the Exploration of the Sea in Copenhagen, March 1931.
Detail from a photograph 15.8 by 23 cm by Pressens Billed Central (Aktuel Billed Central), Royal Library Access. No. 1945–Nr. 63 (negative no. 91099) and a positive 17 by 23.5 cm, Royal Library, Access. No. 1975–882126 (Negative no. 124672). Figure 51 is a reproduction of the last mentioned picture. The photograph was taken in the staircase of the Carlsberg Laboratory on Wednesday 25. March 1931. Details of the photograph with a short text were published in the daily newspapers BERLINGSKE MORGENAVIS 26. March on the back page: Fest for forskerne, and in B.T. the same day on the middle page: En straalende Fest for Havforskerne. After the dinner the guests were shown the Dana-film and the collections in the attic of the Carlsberg Laboratory. In the front row from left to right:
(1) the Danish Prime Minister Th. Stauning,
(2) the hostess Mrs. I. Schmidt,
(3) the French Vicepresident of the Council Mr. T. Tissier, (Paris),
(4) the host Dr. Johs, Schmidt,
(5) the British President of the Council Mr. H. G. Maurice, (London),
(6) Professor Johan Hjort (Oslo),
(7) H.R.H. Prince Valdemar of Denmark,
(8) H.R.H. Prince Georg of Greece.

52 On the occasion of the »Dana«s return an exhibition of ethnographical objects was held at the Carlsberg Laboratory, some of them shown by the laboratory's assistants and friends (1931).
From an original positive 16.8 by 22.8 cm by Thv. Larsen, Fotograf, BERLINGSKE TIDENDE. (By courtesy of Mrs. Vita Holm, née Hansen). This very picture was published in the weekly magazine TIDENS KVINDER 9. Aarg. No. 8, 24. February 1931 page 22, together with two other pictures and text filling one complete page.
The exposure was made either at the dress rehearsal Saturday 7. February (Johs. Schmidt's diary) or at the receptions held on Sunday 15. or Sunday 22. February 1931. The daily newspaper BERLINGSKE TIDENDE reported the last mentioned events on Tuesday 12. and Monday 23. February 1931 but with some other photographs. The receptions were on a grand scale comprising at least 20 people dressed in exotic garments, but also music, dances, slides, films and lectures. The many guests were select such as the Prime Minister Th. Stauning, the Minister of education, scientists etc.
The persons seen are (left to right); Last row (1) Bror van der Aa. Kühle (2) Esther Hansen (3) Peter Anton Koch; First row: (4) Inger Holm (5) Vita Hansen (6) Lise Lützen-Nielsen. (1), (2), (3) & (5) are in Javanese, while (4) & (6) are in Sumatran (Batak) costumes. The complete ethnographical collection afterwards went to the National Museum in Copenhagen.
C.F.P.A. No. 1975–520.

53 Øjvind Winge and his hunting dog, »Lasse« 1956. Detail from a photographic negative 6 by 6 cm by Janne Woldbye (© A/S Pressehuset, Copenhagen; No. 6E3/31A6), taken on the occasion of Ø. Winges seventieth birthday. C.F.P.A. No. 1966–39.

54 One of the Champagne bottle messages. From the Øjvind Winge archives, envelope marked: Vedr.. Sargassum. A 9 by 14 cm waterlogged postcard. C.F.P.A. No. 1976–140, and 1976–141.

55 Winge as Johs. Schmidt's assistant (1913). Detail from a mounted photograph on printing-out paper 6 by 9 cm, (carte-de-visite) by Juncker-Jensen, Copenhagen, 1913. By courtesy of The Botanical Library of the University of Copenhagen. C.F.P.A. No. 1966–192.

56 Winge's drawings of tumour cells. (a) and (b) taken from: Ø. Winge: Zytologische Untersuchungen über die Natur maligner Tumoren. I. »Crown Gall« der Zuckerrübe. Z. Zellforsch. Mikroscop. Anat. *6* p. 405 Abb. 6 & Abb. 5. (1927). Magnification c. 3000 x.
(c) and (d) taken from: Ø. Winge: Zytologische Untersuchungen über die Natur maligner Tumoren. II. Teerkarzinome bei Mäusen. Z. Zellforsch. Mikroscop. Anat. *10*, p. 704 Abb. 21 & p. 705, Abb. 23. Magnification (c) c. 2100 x, (d) c. 1600 x.

57 Winge working with the micro-manipulator (1942). Detail from a mounted and framed photograph 18 by 18 cm by unknown photographer; 22. February 1942. By courtesy of Brock & Michelsen, Copenhagen. C.F.P.A. No. 1964–598.

58 Hybridization of yeast.
(a) to (d) From: Ø. Winge and O. Laustsen: On two types of spore germination, and on genetic segregations in *Saccharomyces*, demonstrated through single-spore cultures. Compt.Rend.Trav.Lab.Carlsberg. Sér.Physiol. *22* p. 102, fig. 1–4 (1940). (Magnification: c. 460 x).
(e) to (f) From: Ø. Winge and O. Laustsen: Artificial species hybridization of yeast. Compt. Rend.Trav.Lab.Carlsberg.Sér.Physiol. *22* p. 247 and Plate I, No. 3 and 4. (1940). (Magnification: c. 840 x).
(a) to (f): C.F.P.A. No. 1975–532.
(G) From: Ø. Winge and Catherine Roberts: Inheritance of enzymatic characters in yeasts, and the phenomenon of long-term adaption. Compt.Rend.Trav.Lab.Carlsberg.Sér.Physiol. *24* p. 315 Plate I (1950). C.F.P.A. No. 1975–533.

59 Niels Nielsen, 1930. Detail from a photograph 15 by 10 cm by Johannes Haurslev, Copenhagen, taken March 1930 (cabinet size, Plate No. 51526) but copied from the negative 1939. By courtesy of the Botanical Library of the University of Copenhagen. C.F.P.A. No. 1976–247.

60 H. Holter, 1964. Detail of a photograph 24 by 18 cm by Hans Petersen, Hornbæk, from 28. May 1964 (36 by 24 mm negative). One out of a series of 24 pictures with H. Holter illustrating a day's work in the laboratory. C.F.P.A. No. 1966–172.

61 Linderstrøm-Lang and Holter during a micro-titration, 1936. The photograph dates back to the weekly magezine, »UGE-JOURNALEN« (»VECKO-JOURNALEN«) Sunday 19. April 1936, No. 16 p. 11. Photographer: unknown. Detail taken from an original copy 24 by 18 cm, at the Carlsberg Museum. C.F.P.A. No. 1964–253.

62 A »Carlsberg constriction pipette«. Taken from: Die Methoden der Fermentforschung (E. Bamann and K. Myrbäck, Editors), Vol I, p. 1139, Fig. 3. Georg Thieme Verlag, Leipzig (1941).

63 The distribution of pepsin in the mucosa of the hog stomach. Both figures taken from Kaj Ulrik Linderstrøm-Lang: Lane Medical Lectures. Proteins and Enzymes. Stanford University Press, Stanford, California (1952). Left: p. 32, Figure 27, Right: p. 34, figure 28.

64 Equipment for Cartesian diver measurements. Left and middle drawings are with slight modifications taken from Kaj Ulrik Linderstrøm-Lang: Lane Medical Lectures. Proteins and Enzymes. Stanford, California (1952). Right drawing (with a scale added) is taken from the catalogue Carlsberg Micro Apparatus Units. Ole Dich, Factory for Mechanical, Electrical Laboratory Apparatus and Scientific equipment. 18, Holmevej, Brøndby Strand, Denmark. C.F.P.A. No. 1976–194.

65 A meeting of the Danish Natural Sciences Research Council (1969). During H. Holter's chairmanship the meetings were held in the assembly room of the Carlsberg Laboratory i.e. 2nd meeting on 25. November 1968 until the 43rd meeting on 28. June 1971.
Detail of a photograph 18 by 24 cm by Bo Jarner (6 by 6 cm negative, © A/S Pressehuset, Copenhagen No. A697/333). The photograph was taken with all members present, at the 5th meeting on Monday 10. February 1969 at 2 p.m. Persons (left to right): (1) P. Olesen Larsen (2) Niels Meyer (3) Tage Carlsen (4) Ben Mottelson (5) Bengt Strömgren (6) Chr. Knakkergaard Møller (7) Heinz Holter (8) P. A. Koch (9) Jytte Kvorning (10) Jørgen Korsbang (11) Chr. Overgaard Nielsen (12) Ove Frydenberg (13) H. H. Ussing (14) A. Noe-Nygaard. C.F.P.A. No. 1969–43.

66 Holter painted by K. Linderstrøm-Lang, about 1933. Oil on canvas 62.8 by 44.4 cm, no inscriptions. By courtesy of Karen and Heinz Holter. Photographed by Hans Petersen, The Royal Museum of Fine Arts, Copenhagen; 12 by 9 cm, colour diapositive. C.F.P.A. No. D75–4.

67 Karen Holter, Heinz Holter and Peter Holter off the island of Strynø in Southern Denmark (1958). Detail of a photographic negative 24 by 36 mm by Erik Zeuthen (summer 1958, No 16–16A). C.F.P.A. No. 1975–514.

68 Members of the early boards of trustees: Paintings in the assembly room at the Carlsberg Laboratory. (The northern wall).
C. T.. Barfoed (1815–1889). Oil on canvas, 97.5 by 73.8 cm by August Jerndorff. Inscribed below on the left: AJ 1886. Has been shown on art exhibitions in Charlottenborg 1887, The Town Hall 1901 and »Forum« 1929; also in Berlin at an international exhibition 1891. A copy, oil on canvas 97 by 73 cm by Povl Jerndorff is situated in the Carlsberg Foundation office. Inscribed below on the left AJ 1886, copie-PovlJ-1929. Another copy, oil on canvas 94.7 by 70.5 cm by Rasmus Christiansen is situated in the Carlsberg Museum (Inventory No. 1506); verso: R x CHR: Kopi 1937 efter Aug. Jerndorf, 1886.
P. L. Panum (1820–1885). Oil on canvas 96.6 by 74.2 cm, by August Jerndorff. Painted posthumously; inscribed below on the right: AJ 1894. Aug. Jerndorff made a copy for the University of Copenhagen, 96.2 by 74.3 cm, inscribed below on the right AJ 1906. (Copenhagen Univ. inventory No. 40).
Japetus Steenstrup (1813–1897). Oil on canvas 96 by 73 cm by August Jerndorff 1887. Inscribed below to the left: AJ 85, copieret 87. The original painting froim 1885 is at The Museum of National History at Frederiksborg Palace, 96 by 72 cm, inscribed below on the left AJ. 85 (Inventory No. A 273).
Eugenius Warming (1841–1924). Oil on canvas, 97 by 78 cm by Knud Larsen. Inscribed below on the left: Knud L 1916.
All four paintings were photographed by: Hans Petersen, The Royal Museum of Fine Arts, Copenhagen, on 12 by 9 cm colour diapositives. C.F.P.A. Nos. D75–7, D75–8, D75–9, and D75–10.

69 Members of later boards of trustees; paintings in the staircase of the Carlsberg Laboratory.
S. M. Jørgensen (1837–1914). Oil on canvas 126 by 100 cm by Herman Vedel. Inscribed below on the left 19 HV 14. This is a copy of the original picture at the Museum of National History at Frederiksborg Palace, oil on canvas 119 by 100 cm, inscribed below on the left HV. 1908. A copy oil on canvas, 97 by 73 cm comprising a detail (only half length) was made by the artist for the Carlsberg Foundation office in 1929; inscribed, verso: HV. The last mentioned picture was copied by Rasmus Christiansen for the Carlsberg Museum

(inventory No. 1503), oil on canvas 95.9 by 70.7 cm; inscribed verso: R x CHR. Kopi 1937 efter H, Vedel 1929.
V. Henriques (1864–1936). Oil on canvas 98 by 78 cm by Hans Henningsen in 1924. Inscribed below on the right: HH.
N. J. Bjerrum (1879–1958). Oil on canvas 104 by 84 cm by Olaf Rude. Inscribed below on the left: Olaf Rude 47.
All three paintings were photographed by: Hans Petersen, The Royal Museum of Fine Arts, Copenhagen; on 12 by 9 cm colour diapositives. C.F.P.A. Nos. D 75–11, D 75–12, and D 75–13.

70 *Knud Jessen* (1884–1971). Detail from a mounted and framed photograph 29 by 24 cm by Jens Juncker-Jensen, Copenhagen, 1928, hanging in the hall of the Carlsberg Laboratory. C.F.P.A. No. 1976–192.

71 *A. C. Andersen* (1882–1970). Detail from a mounted and framed photograph. 14 by 9.9 cm, by Polyfoto, Copenhagen; situated at the Royal Veterinary and Agricultural University, Department of Chemistry. (By courtesy of A. Tovborg Jensen). C.F.P.A. No. 1967–64.

72 *A. Tovborg Jensen.* Detail from a mounted and framed photograph 29 by 23.1 cm, by Elfelt, Copenhagen, 1957; hanging in the hall of the Carlsberg Laboratory. C.F.P.A. No. 1976–191.

73 The Isotope Laboratory, established 1965–66, during the chairmanship of A. Tovborg Jensen, seen from the north-east. Detail of a photographic negative 9 by 12 cm by Strüwing Reklamefoto, Birkerød (©, No. 257/38), springtime 1969. The building was designed by Architect Paul Niepoort and described by him in the bi-monthly periodical »ARKITEKTUR« *13*, No. 5, pp 201–203 (1969). C.F.P.A. No. 1969–186.

74 The present Board of Trustees, 1976. Left to right: A. W. Nielsen, Chr. Crone, Ebba Lund, A. Noe-Nygaard, C. J. Ballhausen. Detail from a photograph (6 by 6 cm negative) taken by Rigmor Mydtskov at a meeting in the assembly room at the Carlsberg Laboratory on 21. January 1976. (Negative No. 31448–7). C.F.P.A. No. 1976–168.

75 The entrance hall of the Carlsberg Laboratory. Two marble busts on square socles are seen. The left one, depicting Louis Pasteur (70 cm high) is by Paul Dubois. Inscribed on the left side of the bust (the viewer's right): P. Dubois. Commissioned by J. C. Jacobsen for the Carlsberg Laboratory in 1878. Several copies in plaster and bronze are known, f.ex. at the Carlsberg Museum, the Carlsberg Breweries and Ny Carlsberg Glyptothek.
The right bust: Justus von Liebig (74 cm high) was possibly made by Michael

Wagmüller (a copy; year: ?, no inscriptions). Photographed by Bloch & Claussen A/S, 1937, detail from a 9 by 12 cm plate negative (B. C. 11938b) (© Teknisk Foto A/S, Copenhagen).

76 Henry Lanz, between experiments.
Taken from a photograph 9.5 by 14.3 cm, by an unknown photographer (around December 1936).
C.F.P.A. No. 1963–67.

77 The Lunchtable, 1952.
Detail from a photograph (6 by 5 cm negative) by Daniel Steinberg. Persons, left to right:
(1) Barbara Wright. (2) Gertie Perlmann. (3) Carsten Olsen. (4) K. U. Linderstrøm-Lang. (5) Halvor N. Christensen. (6) Martin Ottesen. (7) Ruth Hubbard. (8) Jytte Hjo Taplov. (9) Arne Johansen. (10) to the utmost left, the back of Bodil Meedom (née Dahlerup Petersen).
By courtesy of Jytte Hjo Taplov, Roskilde.
C.F.P.A. No. 1975–200

78 What S. P. L. Sørensen politely did not see: R. D. Hotchkiss and H. Holter, drawn by Linderstrøm-Lang on the blackboard of the Laboratory.
Taken from a photograph 7.9 by 5.4 cm, by an unknown photographer sometime between 1937 and 1938.
C.F.P.A. No. 1963–178.

79 Carlsberg reunion in Brussels, 1955.
In connection with the 3rd International Congress of Biochemistry in Brussels a Carlsberg reunion dinner was held at an Italian restaurant on Thursday 4. August 1955. About 43 alumni were present, of which this photograph shows 30 persons. Recognisable are from left to right:
1. I. Leslie, 2. D. B. Wetlaufer, 3. L. Wetlaufer, 4. E. Fredericq, 5. I. Harris, 6. J. Vandeputte, 7. H. Holter, 8. Aa. Vaslow, 9. F. Vaslow, 10. C. Schellman, 11. K. Nilsson, 12. Mrs Murray Luck, 13. B. Max Møller, 14. J. Murray Luck, 15. L. Vandendriessche, 16. H. Tuppy, 17. A. Wollenberger, 18. A. Bronfenbrenner, 19. T. Rosenberg, 20. K. U. Linderstrøm-Lang, 21. G. Linderstrøm-Lang, 22. H. Neurath, 23. G. Perlmann, 24. H. L. Fraenkel-Conrat, 25. J. Fraenkel-Conrat, 26. Mrs. J. Leonis, 27. J. E. Leonis. Taken from a photograph 7.3 by 12.4 cm by an unknown professional photographer.
C.F.P.A. No. 1963–152 C.

80 L. Vandendriessche, drawn by Linderstrøm-Lang for the Christmas party 1947.
Detail of a drawing, crayon on cheap paper (originally a roll 94 cm wide). This drawing is deposited together with all other Linderstrøm-Lang dra-

wings in the Royal Library, Department of Maps and Pictures in Copenhagen as No. 74. Photographed by the Royal Library, (negative no. 85656). C.F.P.A. No. 1964–328.

81 How the colloquia were advertised. The Carlsberg Laboratory is not mentioned; everybody knew the small yellow sheets of foolscap, and where the colloquia were held.
From a genuine note board slip 9.8 by 14.8 cm, made by typewriting the message upon a spirit duplicator master paper in violet and running the master with the standard small sheetes of yellow foolscap for about fifty copies. They were mailed about a week in advance. The Standard small sheets of yellow foolscap were otherwise used all over the Carlsberg Laboratory as scribbling paper, though varying a little in size over the years from S. P. L. Sørensen's time until now.

82 Mushroom hunting in Jægerspris forest.
Photograph by K. Max Møller, Copenhagen. 26. September 1955. Detail of a 24 by 36 mm negative.
C. F. P. A. No. 1976–56.
Persons (from left to right):
Ernstl Hagens, Wolfgang Hagens, (Heinz Holter; only hat, legs and map visible), Kaj Hjelm (his back), Hellen Ziegler, Karen Holter, Birte Max Møller, Eva Kieler (half covered) and Jørgen Kieler.

83 Johs. Schmidt meets in 1929 on the island of Koh Chang (Thailand) his old collaborator Nai Sug, who 30 years earlier on Schmidt's very first expedition had assisted him in his studies of the Thai flora.
This happened during »Dana«'s circumnavigation of the globe 1928–1930. Detail from a film negative 10 by 8 cm by an unknown photographer. Found in the »Dana« collections at »Charlottenlund Slot«.
The last section of a rather long entry in Johs. Schmidt's diary for Wednesday 8. May 1929 tells (translated literally from the Danish): » . . .I was photographed together with my old helper whose name is Sug, now: Nai Sug Sugsai, Koh Chang, Changwad, Chantaburi, Siam, and I will of course send him the picture. He did not look old, in spite of beeing over fifty years he had no gray hair, though the hair was somewhat thin in the forehead and his one leg was not strong. He had never in all those thirty years left Kho Chang! He had several sons. In the village there was now a Wat«.
C.F.P.A. No. 1976–31.

84 Route of the schooner »Margrethe«, 1913.
According to Johs. Schmidt but taken from Poul Jespersen's article (p. 61) in:

Øjvind Winge & Å. Vedel Tåning (Editors): *Naturforskeren Johannes Schmidt.* København: Gyldendalske Boghandel. Nordisk Forlag (1947). Artist: unknown.
C.F.P.A. No. 1976–540.

85 Route of the motor schooner »Dana I«, 1920. Insert: »Dana I«, 1921. According to J. Schmidt but taken from Poul Jespersen's article (p. 82) in the same book as fig. 84. Artist: unknown.
C.F.P.A. No. 1975–541.

86 Route of »Dana II«, 1921–22.
According to Johs. Schmidt but taken from Poul Jespersen & Å. Vedel Tåning's article (p. 90) in the same book as fig. 84. Artist: unknown.
C.F.P.A. No. 1975–542.

87 Breeding grounds of the European eel (Sargasso Sea) and the Sumatra eels.
Taken from Anton Fr. Bruun's article (p. 162) in the same book as fig. 84. Artist: unknown.
C.F.P.A. No. 1975–534.

88 Route of »Dana« II's circumnavigation of the globe 1928–1930. According to Johs. Schmidt. but taken from p. 183 in:
Torben Wolff. *Danish Expeditions on the Seven Seas.*
RHODOS International Science and Art Publishers Copenhagen (1967).
Artist: unknown.
C.F.P.A. No. 1975–535.

89 »Dana« II entering the harbour of Copenhagen after the expedition 30. August 1921 – 11. July 1922.
Taken from a photographic negative 5 by 7.2 cm on a 9 by 12 cm plate. It is a rephotograph of a positive that appeared in the morning newspaper POLITIKEN on Wednesday, 12. July 1922. Page 1. The plate was found in »Charlottenlund Slot«.
C.F.P.A. No. 1976–43.
The ship was originally built in 1917 by Cochrane & Sons Ltd. in Selby, and named »John Quilliam«. It served during the First World War as a minesweeper. In the early part of 1921 the Danish State bought the ship from the British Admiralty in order to replace the fishery-investigation ship »Thor«. The small coal-fuelled steam ship, 360 tons, 42 m in length, 7.25 m in the beam and with a maximum speed of 8 knots, was fitted out as a modern marine research vessel, f.ex. with a special winch with a 10000 m's trawl cable, and given the name »Dana« (II). For the 1928–1930 expedition it got even more modern equipment. »Dana« II served until 22. June 1935,

when at about 6. a.m. in a dense fog it was run down in the North Sea (55°55'N, 7°E) by a German trawler, H. G. 92. »Pickhuben« from Cuxhafen, that also saved all the 22 crew members and brought them to Esbjerg.

90 The »Carlsberg Ridge«.
Taken from p. 28 in: CARLSBERG BLADET November 10th. 1972. 38th volume-No. 5: Carlsberg 125 years.
Artist unknown, but here slightly redrawn by B. Samson Carlsen. C.F.P.A. No. 1975–536.

91 Dana collections.
Thousands of samples from the expedition 1921–22 in the attic of the Carlsberg Laboratory.
Detail from a photographic negative plate 9 by 12 cm by an unknown photographer. Found in the »Dana« collections at »Charlottenlund Slot«. C.F.P.A. No. 1976–267.

92 S. A. van der Aa Kühle, 1849–1906.
Detail from a »cabinet« size photograph 16.1 by 10.5 cm by Marius Christensen, Atelier-Moderne, Copenhagen, sometime between October 1899 and April 1906. Here taken from a rephotograph by Bloch & Claussen A/S, Copenhagen (Negative, 12 by 9 cm, No. 12803–7) now owned by Teknisk Foto, Copenhagen.
C.F.P.A. No. 1976–352.

93 Kühle's hand-sketch of the yeast propagating apparatus, and a drawing showing its final shape (compare fig. 43). d: Wort container; c: Propagation cylinder; b: manual stirring device; g-h: entrance of air supply (through filter g).
The left sketch 30 by 20 cm is in the possession of the Carlsberg Breweries. By courtesy of B. Trolle. Photographed by Foto-Service I/S, Copenhagen, Negative No. M 3319. Published before in BRYGMESTEREN; *29*, No. 11 p. 289, Fig. 19 (1972).
C.F.P.A. No. 1976–54.
The right figure is taken from a woodcut 16.6 by 11,2 cm, by J. F. Rosenstand, as published in: Meddelelser fra Carlsberg Laboratoriet *2* (1883–1888) p. 287, Fig. 6, and in the French résumé p. 183, Fig. 3. (1888).
C.F.P.A. No. 1976–353.

94 N. Hjelte Claussen (1866–1955).
From a mounted and framed photograph 29.1 by 24.3 cm by Jens Juncker-Jensen, Copenhagen. Inscribed below on the left Jens Juncker Jensen, 1938. On *verso:* 27. April 1938; (but it is possibly taken 8. July 1936; Negative No. 272528). Situated in the hall of the Carlsberg Laboratory.
C.F.P.A. No. 1976–190.

95 Building of the Research Laboratory of the Carlsberg Breweries, now »Central Laboratory«.
This composite structure is situated on a triangular site between Ny Carlsbergvej and Kammasvej. The somewhat pompous building in the center and the left of the picture was originally built as main office building for Carl Jacobsen by the architect Hack Kampmann in 1901. It was built onto the villa (from 1873) seen in the right side of the picture. The very low extension on the lower right side, contained the The Svedberg ultracentrifuge (1950–1974), but is now (1976) being replaced by a large modern building. To the far left, in the background, a 87 m high silo tower is under construction. Detail of a photographic negative 6 by 6 cm by F. Aggersbo (© Foto-Service I/S, Copenhagen, Nr. J2/1867) taken on 1. May 1963.
C.F.P.A. No. 1976–15.

96 Harvest of hops at the old experimental farm »Nordgården«. Detail from a photograph 9 by 12 cm negative by F. Aggersbo (© Foto-Service I/S, No. u4470) taken on 24. October 1953.
C.F.P.A. No. 1976–16.

97 The present experimental farm »Allindemaglegård« on Central Sealand, near Ringsted. In the foreground the many experimental plots for barley.
From a photograph 24.2 by 30.4 cm by Knud Brethvad, Virum (CB. 1794) taken in springtime 1972. Printed before in BRYGMESTEREN *29*, No. 11. p. 270 (1972).
By courtesy of the Carlsberg Breweries, Department of Information.
C.F.P.A. No. 1976–205.

98 Expenditures of the Carlsberg Foundation during the years 1961–1971.
Original figure by P. Brandt Rehberg, drawn by B. Samson Carlsen.
C.F.P.A. No. 1975–537.

99 Diter von Wettstein working at the electron microscope. Detail of a photographic negative 36 by 24 mm by Rigmor Mydtskov, Copenhagen (Negative No. 31456(13–14) taken 27. March 1976 at the Carlsberg Laboratory). The microscope is a »Siemens Elmiskop 102« with image intensifier and scanning transmission attachment.
C.F.P.A. No. 1976–240 (detail of No. 1976–66).

100 (100,a) Dark grown leaves of wild type barley (right) and the regulatory mutant *tigrina-d*^{12} which accumulates large amounts of the green pigment protochlorophyllide, a precursor to chlorophyll.
(100,b) Progeny segregating for seedlings of the mutant *tigrina-d*^{12} (short),

seedlings heterozygous for this mutant gene (bent tip) and wild-type seedlings. The plants are grown in alternating light and dark cycles.
(100,c) Under continuous light in a phytotron the conditional lethal mutant *tigrina-d*^{12} is fully viable.
(100,d) Leaves of the structural gene mutant *xantha-l*^{35}. The leaves on the left have been placed in buffer solution, those on right were provided with δ-aminolevulinic acid. The latter leaves have accumulated the reddish Mg-protoporphyrin.
All four are colour diapositives 36 by 24 mm by Poul Eriksen, Institute of Genetics, University of Copenhagen. By courtesy of Diter von Wettstein.

101 Development of an etioplast into a chloroplast···.
From original photographs by courtesy of K. W. Henningsen, J. E. Boynton, O. F. Nielsen, and D. von Wettstein. Also published in: Ruth Sager: *Cytoplasmic Genes and Organelles.* Academic Press, New York & London (1972) as Plate XV.
(a) Neg. No. 9082 N; magnification 35,000 x
(b) Neg. No. 8170 N; magnification 46,000 x
(c) Neg. No. Z 4475 H; magnification 17,000 x
(d) Neg. No. 17549 H; magnification 38,000 x
(e) Neg. No. 7255 N;magnification 125,000 x

102 Plastid membrane organizations found in five different nuclear gene mutants of barley···.
From original photographs by courtesy of K. W. Henningsen, J. E. Boynton, O. F. Nielsen, and D. von Wettstein. Also published in Ruth Sager: *Cytoplasmic Genes and Organelles.* Academic Press, New York & London (1972) as Plate XVI.
(a) Neg. No. Z 5139 H; magnification 36,000 x
(b) Neg. No. 9151 H; magnification 70,000 x
(c) Neg. No. 10190 H; magnification 30,000 x
(d) Neg. No. Z 9697 N; magnification 84,000 x
(e) Neg. No. 5914 N; magnification 50,000 x

103 Ground plan of the Carlsberg Research Center.
From a drawing c. 14 by 25 cm by architect Jørgen Buschardt, Copenhagen.
C.F.P.A. No. 1976–354.

104 The Carlsberg Research Center during construction, late 1975. Detail of a colour diapositive 6 by 6 cm by Jakob Bonfils, Copenhagen (August-September 1975). Taken from the top of a building crane.
C.F.P.A. No. D75–14.

105 The Carlsberg Laboratory 1976.
Detail of a colour diapositive 6 by 6 cm by Jakob Bonfils, Copenhagen (18. March 1976). In the middle is seen a bronze statue of J. C. Jacobsen (265 cm) by Vilhelm Bissen. It was unveiled at the inauguration of the building on 10. November 1897. A study (plaster statuette, 30 cm. inventory No. 2297) is kept at the Museum of National History at Frederiksborg Palace.
C.F.P.A. No. D76-3.

106 (pp. 446–447)
Map of Carlsberg territory.
The top half of a map 18.3 by 13.7 cm (scale 1:4000) by architect M. K. Michaelsen published in ARKITEKTEN, vol. *31*, p. 145 (1929). Here slightly redrawn by B. Samson Carlsen.
The numbers on the buildings refer to the year of construction.
C.F.P.A. No. 1976–355.

The Table on page 209:
Johannes Schmidt's expeditions, was found in the »Dana« Archives of The Danish National Archives (No. 10291) The compiler: unknown.

PORTRAITS OF CONTRIBUTORS IN CHAPTER XIII, PP. 397–404.

Lise Allen (née Lützen-Nielsen). Detail of photo: Fritz Kulsbjerg Larsen, Lyngby (neg. 12 by 9 cm; 1969). C.F.P.A. No. 1976–65.

Ronald P. Bell Detail of photo: The Audiovisual Aids Department. University of Stirling, Scotland, (pos. 20.4 by 15.3 cm; 1975). C.F.P.A. No. 1976–307.

Christian Crone Detail of photo: Rigmor Mydtskov, Copenhagen (neg. 6 by 6 cm No. 31453–8; 21. January 1976). C.F.P.A. No. 1976–157.

Vilhelm Ege Detail of photo: unknown photographer and date (pos. 9.9 by 6.9 cm). By courtesy of Mrs. Vita Holm. C.F.P.A. No. 1975–517.

Bent Foltmann Detail of photo: Jytte Hjo Taplov (neg. 6 by 6 cm; September 1961). C.F.P.A. No. 1975–524.

David Glick Detail of photo: unknown photographer (pos. 20.1 by 15.0 cm; 1951). C.F.P.A. No. 1961–28.

Heinz Holter Detail of photo: Rigmor Mydtskov, Copenhagen (neg. 6 by 6 cm No. 31474–7; 27. January 1976). C.F.P.A. No. 1976–308.

Karen Holter (née Teisen). Detail of photo: F. Aggersbo, Foto-Service I/S, Copenhagen (neg. 6 by 6 cm No. H2/34 42b; 10. October 1962). C.F.P.A. No. 1976–69.

Rollin D. Hotchkiss Detail of photo: unknown photographer and date (pos. 24.1 by 17.7 cm). C.F.P.A. No. 1961–38.

A. Tovborg Jensen Detail of photo: Ulf Nellemose, Frederikssund (neg. 6 by 6 cm) of a bronze bust 54 cm high by Knud Nellemose, unveiled in the staircase hall of the Carlsberg Laboratory on 22. February 1974. C.F.P.A. No. 1976–61.

Wilh. Johannsen Detail of photo: A. Lønborg, Copenhagen (pos. carte-de-visite, 94 by 61 mm). Inscribed on *verso:* W. Johannsen, 1889 (En af Dr Salomonsens mest taknemmelige Elever). By courtesy of the Museum of Medical History, University of Copenhagen). C.F.P.A. No. 1969–156.

Herman M. Kalckar Detail of photo: Rigmor Mydtskov, Copenhagen (neg. 6 by 6 cm, No. 29313–2; 1974) C.F.P.A. No. 1976–309.

Jørgen Kieler Detail of photo: unknown photographer and date (pos. 62 by 45 mm). By courtesy of the Danish Cancer Society. C.F.P.A. No. 1976–51.

Albert Klöcker Detail of photo: probably Fred. Riise, Copenhagen. (pos. elliptic, 9 by 6 cm, part of a composite, mounted and framed picture of assistants, by Fred. Riise; 1901). C.F.P.A. No. 1967–1 (part of No. 1901–1).

Rudolph G. Koefoed Detail of photo: N. Nyberg, Copenhagen (pos. cabinet size, 14.5 by 10.0 cm) no date. By courtesy of The Royal Library, Copenhagen, Department of Maps and Pictures (no. 1917–713/98; Neg. No. 138857). C.F.P.A. No. 1975–483.

Poul Larsen Detail of photo: Elfelt, Copenhagen (pos. 14.5 by 10.5 cm; 25. October 1969). By courtesy of Poul Larsen, Aarhus. C.F.P.A. No. 1976–320.

K. Linderstrøm-Lang Detail of photo: Børge Walther, Copenhagen. (pos. 24 by 25.5 cm; 1957). By courtesy of »SE & HØR«, Carl Allers Etablissement A/S, Copenhagen. C.F.P.A. No. 1964–252.

K. Max Møller Detail of photo: Rigmor Mydtskov, Copenhagen (neg. 6 by 6 cm, No. 31477–4; 27. January 1976). C.F.P.A. No. 1976–310.

Detlev Müller Detail of photo: Albert Schou, Jun. Eftf., Copenhagen (pos. 17.9 by 13 cm; 1965). By courtesy of Erika Löhr, Institute of Plant Physiology, University of Copenhagen. C.F.P.A. No. 1976–311.

E. Steemann Nielsen Detail of photo: Lind Studio, Virum (pos. 11.2 by 8.1 cm; 1974). By courtesy of E. Steemann Nielsen. C.F.P.A. No. 1976–312.

P. Brandt Rehberg Detail of photo: Steen Jacobsen, Nordisk Pressefoto A/S, Copenhagen (pos. 18.2 by 11.9 cm; 29. May 1969).
C.F.P.A. No. 1976–319.

Charlotte Schellman (née Green). Detail of photo: Oscar Palmquist (U.S.A.), (pos. 23.4 by 15.7 cm; 1975). By courtesy of Charlotte Schellman, Eugene, Oregon, U.S.A. C.F.P.A. No. 1976–313.

John A. Schellman Detail of photo: Oswald Pfenninger (U.S.A.), (pos. 8 by 11.7 cm; 1974). By courtesy of Charlotte Schellman, Eugene, Oregon, U.S.A. C.F.P.A. No. 1976–314.

Holger Schiönning Detail of photo: probably Fred. Riise, Copenhagen (pos. elliptic, 9 by 5.8 cm,part of a composite, mounted and framed picture of assistants by Fred. Riise; 1901). C.F.P.A. No. 1975–421 (part of No. 1901–1).

Birger Trolle Detail of photo: Janne Woldbye, A/S Pressehuset, Copenhagen (neg. 6 by 6 cm, No. R635/173; 9. August 1963). By courtesy of B. Trolle, Central Laboratory, The Carlsberg Breweries. C.F.P.A. No. 1976–315.

L. Vandendriessche Detail of photo: NIEUWE GAZET (Belgium), (pos. 6.2 by 5.1 cm; 1973). By courtesy of L. Vandendriessche, Dept. Biochemistry, University of Ghent, Belgium. C.F.P.A. No. 1976–316.

Mogens Westergaard Detail of photo: Ebbe Andersen, ©POLITIKENS Pressefoto, Copenhagen. (pos. 23.5 by 18 cm; October 1965). C.F.P.A. No. 1975–151.

Diter von Wettstein Detail of photo: Rigmor Mydtskov, Copenhagen (neg, 6 by 6 cm; 21. January 1976 at the Carlsberg Laboratory). C.F.P.A. No. 1976–317.

Erik Zeuthen Detail of photo: J. Juncker-Jensen, Copenhagen (pos. 18 by 13 cm (neg. 12 by 9 cm, No. 23042–L); 1969). By courtesy of Erik Zeuthen, Biological Institute of the Carlsberg Foundation. C.F.P.A. No. 1976–318.

Index

The index includes only names, not subjects. If the full christian names of persons could not be ascertained, only initials are given. For persons with several christian names the one most commonly used has been marked by italics. The English alphabet has been used throughout, so that the letters æ, ø, ö and ü are found under ae, oe and ue.

Names of institutions were only included, if they are characterized by a personal name instead of a country or city. All institutions connected to »Carlsberg« have been omitted.

Page numbers referring to illustrations are marked by italics. Cross-references will be found but are restricted to a minimum.

L

X-Y

Z

106 Map of Carlsberg territory

A. J. C. Jacobsen's villa at Gamle Carlsberg (now »Mansion of Honour«).
B. The first Carlsberg Laboratory.
C. Building in which the oldest Brewery Laboratory was located.
D. Gamle Carlsberg Brewery Laboratory of 1886.
E. Greenhouse of the Carlsberg Laboratory (1911).
F. Living quarters of head of chemical department.
G. Carlsberg Laboratory (1897), Gl. Carlsbergvej 10.
H. Living quarters of head of physiological department.
I. Subdepartments of fermentation physiology and cytochemistry, Gl. Carlsbergvej 8.
K. Present Brewery Laboratory (»Central Laboratory«), from 1882 to 1906 the site of the Ny Carlsberg Brewery Laboratory.
L. The first Ny Carlsberg Brewery, later known as »The Middle Brewery«
M. Site of the original Ny Carlsberg Brewery Laboratory.

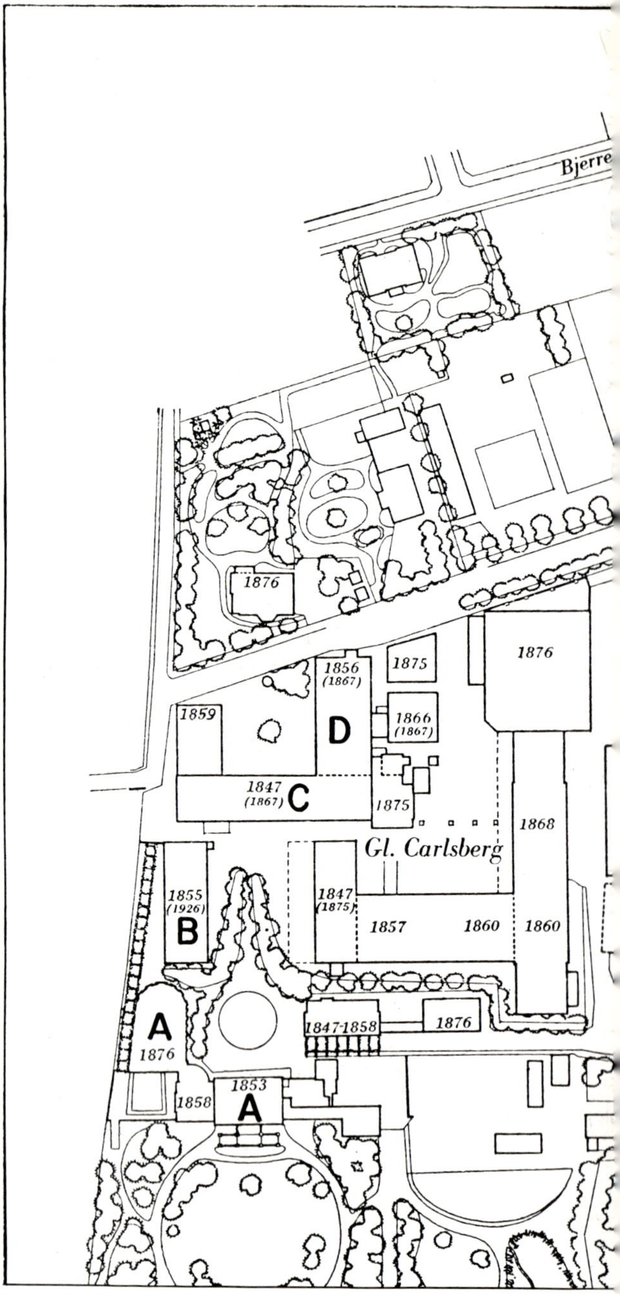

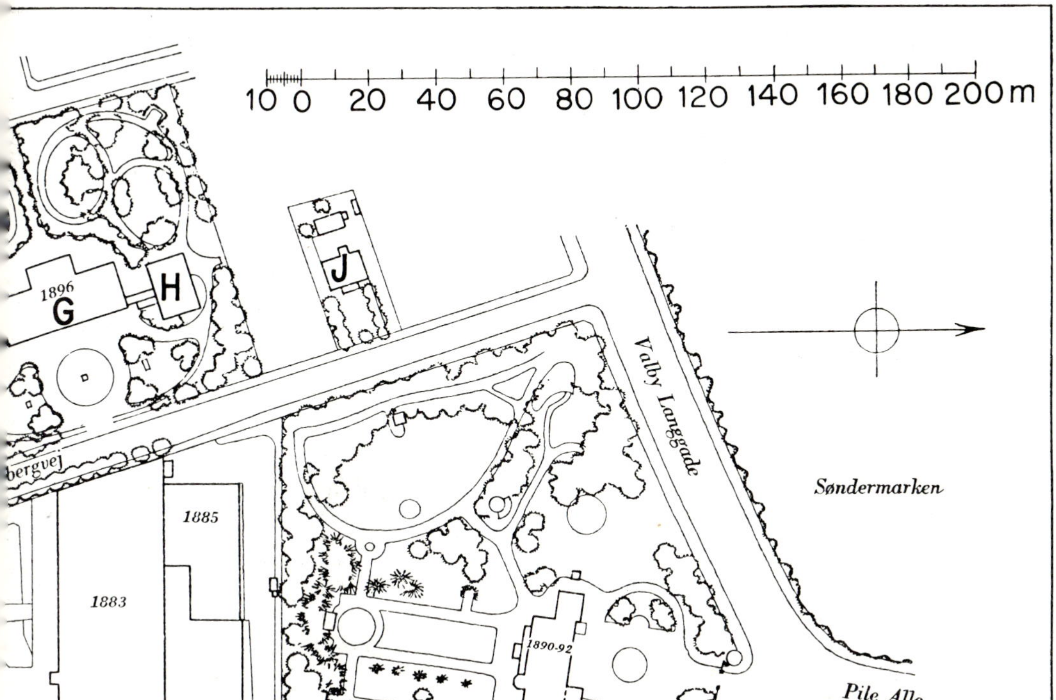
10 0 20 40 60 80 100 120 140 160 180 200 m
1896
G
H
J
Valby Langgade
bergvej
Søndermarken
1885
1883
1890-92
Pile Alle
1893
1894
1873
1877
L
1896
Kammasvej
1883
llembryggeriet
K
1901
1880
L
1873
1876
Ny Carlsberg
1901
1901
1882/1928
1927
1883
1901
1920
1881